2nd International Conference on Materials Engineering and Applications (ICMEA 2019)

IOP Conference Series: Materials Science and Engineering
Volume 540

Da Nang, Vietnam
16 – 18 January 2019

ISBN: 978-1-5108-8920-0
ISSN: 1757-8981

Printed from e-media with permission by:

Curran Associates, Inc.
57 Morehouse Lane
Red Hook, NY 12571

Some format issues inherent in the e-media version may also appear in this print version.

Printed by Curran Associates, Inc. (2019)

For permission requests, please contact the Institute of Physics at the address below.

Institute of Physics
Dirac House, Temple Back
Bristol BS1 6BE UK

Phone: 44 1 17 929 7481
Fax: 44 1 17 920 0979

techtracking@iop.org

Additional copies of this publication are available from:

Curran Associates, Inc.
57 Morehouse Lane
Red Hook, NY 12571 USA
Phone: 845-758-0400
Fax: 845-758-2633
Email: curran@proceedings.com
Web: www.proceedings.com

Table of contents

Volume 540

2019 2nd International Conference on Materials Engineering and Applications (ICMEA 2019)

16–18 January 2019, Da Nang, Vietnam

Accepted papers received: 23 April 2019

Preface

Papers

Material Physics and Thin Film Technology

Material Chemistry and Chemical Engineering

Metal Processing and Material Forming

Material Design and Material Mechanics

IOP Conf. Series: Materials Science and Engineering **540** (2019) 011001 doi:10.1088/1757-899X/540/1/011001

PREFACE

The 2nd International Conference on Materials Engineering and Applications 2019 (ICMEA 2019) was held in Da Nang, Vietnam during January 16-18, 2019.
ICMEA 2019 had a great success to collect many researchers from various countries, and its academic atmosphere helped participants to enjoy scientific and technological discussions for the development of advanced materials in the field of fundamental and applied materials sciences.

ICMEA 2019 provided an entirely real platform for researchers, engineers and academicians to further explore materials engineering and applications. Distinguished professors were also invited to deliver keynote speeches relating to the latest developments and trends in their respective fields, including nano/micro/macro-scale materials sciences and technologies. It is a golden opportunity for students, early-career researchers and engineers to meet the experts and specialists and get their advice or consultation on technical matters, sales and marketing strategies.

This set of conference proceedings presents a selection of papers submitted to the conference from various universities, research institutes and industries. All papers were subjected to peer-review by conference committee members and other international reviewers. The papers were selected according to high standards for technical accuracy, scientific rigor, and presentation quality in this conference. This volume presents recent advances in the field of Materials Science and Engineering, Materials Processing Technology, Materials Properties, Measuring Methods and Applications, etc.

We sincerely thank all the authors who have contributed to this volume and also to the organizing committee, reviewers, speakers, chairpersons, sponsors and all the conference participants for their significant support to ICMEA 2019.

Prof. Takashige Omatsu,
Chiba University, Japan

Conference Co-chairs

Prof. Stephen John Pennycook, National University of Singapore, Singapore
Prof. Shaojun Guo, Peking University, China
Prof. Takashige Omatsu, Chiba University, Japan
Prof. Jae-Jin Shim, Yeungnam University, South Korea

Program Chairs

Prof. Sheikh Akbar, The Ohio State University in Columbus, USA
Prof. Frank Otremba, Federal Institute for Material Research and Testing, Germany

Technical Committee

Prof. Cengiz S Ozkan, University of California, USA
Prof. Yalin Dong, University of Akron, USA
Prof. Nicola Pugno, University of Trento, Italy
Prof. Guy Le Lay, Aix-Marseille Université, France
Prof. Zhongchun Chen, Tottori University, Japan
Prof. Khalil El-Hami, University of Hassan 1st, Morocco
Prof. Haiqing Yin, University of Science & Technology Beijing, China
Prof. Yanqing Yang, Northwestern Polytechnical University, China
Prof. Yadong Xu, Northwestern Polytechnical University, China
Prof. Yuanzhi Zhu, North China University of Technology, China
Prof. Yanjing Su, University of Science & Technology Beijing, China
Prof. Chul Hee Lee, Inha university, South Korea
Prof. Jae-Jin Shim, Yeungnam University, South Korea
Prof. O.L. Shanmugasundaram, K.S. Rangasamy College of Technology, India
Prof. Y. Sangappa, Mangalore University, India
Prof. N. Ethiraj, Dr. M.G.R Educational and Research Institute-University, India
Prof. Hosen Ali Jawan, Univeristy Of Elmergib, Libya
Prof. Sumit Srivastava, Manipal University Jaipur, India
Prof. Yarramaneni Sridharbabu, BML Munjal University, India
Assoc. Prof. Daniel H. C. Ong, Chinese University of Hong Kong, Hong Kong
Assoc. Prof. Chengliang Hu, Shanghai Jiao Tong University, China
Assoc. Prof. Xifeng Li, Shanghai Jiao Tong University, China
Assoc. Prof. Sha Shiong Ng, Universiti Sains Malaysia, Malaysia
Assoc. Prof. Sushil Kumar Jain, Manipal University Jaipur, India
Assoc. Prof. Madan Mohan Sharma, Manipal University Jaipur, India
Assoc. Prof. Sanyog Rawat, Manipal University Jaipur, India
Dr. Xiaoliang Wei, Pacific Northwest National Laboratory, USA
Dr. Andrzej Grzechnik, RWTH Aachen University, Germany
Dr. Raj Das, The University of Auckland, New Zealand
Dr. Alain Kusmoko, University of Wollongong, Australia

IOP Conf. Series: Materials Science and Engineering **540** (2019) 011002 doi:10.1088/1757-899X/540/1/011002

IOP Conf. Series: Materials Science and Engineering **540** (2019) 011003 doi:10.1088/1757-899X/540/1/011003

Peer review statement

All papers published in this volume of *IOP Conference Series: Materials Science and Engineering* have been peer reviewed through processes administered by the proceedings Editors. Reviews were conducted by expert referees to the professional and scientific standards expected of a proceedings journal published by IOP Publishing.

IOP Conf. Series: Materials Science and Engineering **540** (2019) 012001 doi:10.1088/1757-899X/540/1/012001

Energy Transfer between Ce^{3+}-Dy^{3+} In $Ca_2Al_2SiO_7$: Ce^{3+}, Dy^{3+} Phosphor

Do Thanh Tien[1,2], Nguyen Manh Son[2], Le Van Tuat[2], Le Ngoc Liem[3]

1 Faculty of Basic Science, University of Agriculture and Forestry, Hue University, Vietnam
2 Faculty of Physics, University of Science, Hue University, Vietnam
3 Office of the Board of Trustees, Duy Tan University, Vietnam
E-mail: dothanhtien@huaf.edu.vn

Abstract: Ce^{3+} and Dy^{3+} ions doped $Ca_2Al_2SiO_7$ (CAS) phosphors were prepared by the solid-state reaction at 1280^0C for 1h. X-ray diffraction patterns confirmed a tetragonal crystalline structure. The luminescent spectra of CAS: Ce^{3+}, Dy^{3+} phosphors consist a board band with peaking at 420 nm corresponding to the luminescence of Ce^{3+} ions and narrow lines of Dy^{3+} ions. The energy transfer from Ce^{3+} ion to Dy^{3+} ion is presented and discussed.

1. Introduction

In lighting and display technology, luminescent material has an important role in manufacturing fluorescent lamp and LED which are highly efficient and energy saving. Rare earth doped luminescent materials are widely used in various applications because they are non-toxic, environmental friendly, and have high luminosity and life expectancy[1]. In recent years, white LED was excited by near ultra violet radiation that is combined with red, green and blue luminescent materials have attracted attention from many scientists. Materials that emit visible light with high luminescent efficiency under stimulation by near UV radiation or blue light have been applied in white LED manufacturing. Alkaline earth alumino silicate materials have attracted a lots of attention from scientist, and become an interesting research topic in the luminescent material field because their high chemical stability and better water resistance compared to the materials on sulfite and aluminate lattice[2-5]. This paper presents research results on luminescence and energy transfer from Ce^{3+} ion to Dy^{3+} ion in $Ca_2Al_2SiO_7$ lattice.

2. Experiments

Ce^{3+}, Dy^{3+} ion doped $Ca_2Al_2SiO_7$ (CAS) phosphor were prepared by the solid-state reaction. Chemical reagents include: $CaCO_3$ (99.9%, China), Al_2O_3 (99%, China), SiO_2 (99.9%, Korea) and CeO_2 (99.9%, Merck), Dy_2O_3 (99.9%, Merck). These precursors are weighted by molar ratio and mixed with 4 weight percent B_2O_3 (used as fluxing agents). The mixtures are well grinded for two hours by using an agate mortar. Then the mixtures were annealed at 1280 °C for 1 h. The X-ray diffraction (XRD) diagram was measured by using the diffractometer Bruker D8-Advance. Photoluminescence (PL) and Photoluminescence excitation (PLE) spectra were measured by Fluorescence spectrometer FL3-22 Horiba with 450W Xenon lamp.

3. Results and discussions

3.1. Crystal structure of single doped and co-doped CAS

IOP Conf. Series: Materials Science and Engineering **540** (2019) 012001 doi:10.1088/1757-899X/540/1/012001

The analysis results show that the samples have tetragonal phase structure, particularly $Ca_2Al_2SiO_7$ with

lattice parameters: a = b = 7.69000 $\overset{0}{A}$, c = 5.06750 $\overset{0}{A}$, α = β = γ = 90°, and space group P-421m. The crystalline structure of the single-doped sample (CAS: Ce^{3+} (0.5 mol%), CAS: Dy^{3+} (1 mol%)) were compared to that of co-doped samples (CAS: Ce^{3+} (0.5 mol%), Dy^{3+} (1 mol%)) as shown in Figure 1. The diffraction diagrams do not show any characteristic peaks of rare earth ions as well as of initial reagent components.

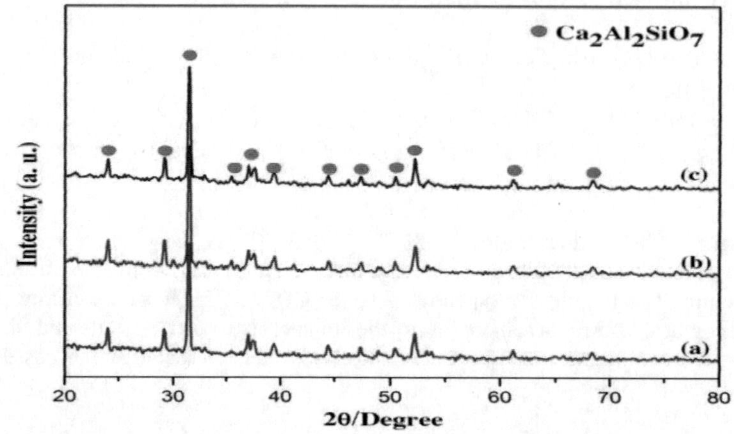

Figure 1. The XRD diagrams of CAS: Dy^{3+} (1 mol%) (a), CAS: Ce^{3+} (0.5 mol%) (b) and CAS: Ce^{3+} (0.5 mol%), Dy^{3+} (1 mol%) (c).

3.2. Energy transfer between Ce^{3+}, Dy^{3+} in the $Ca_2Al_2SiO_7$ phosphor

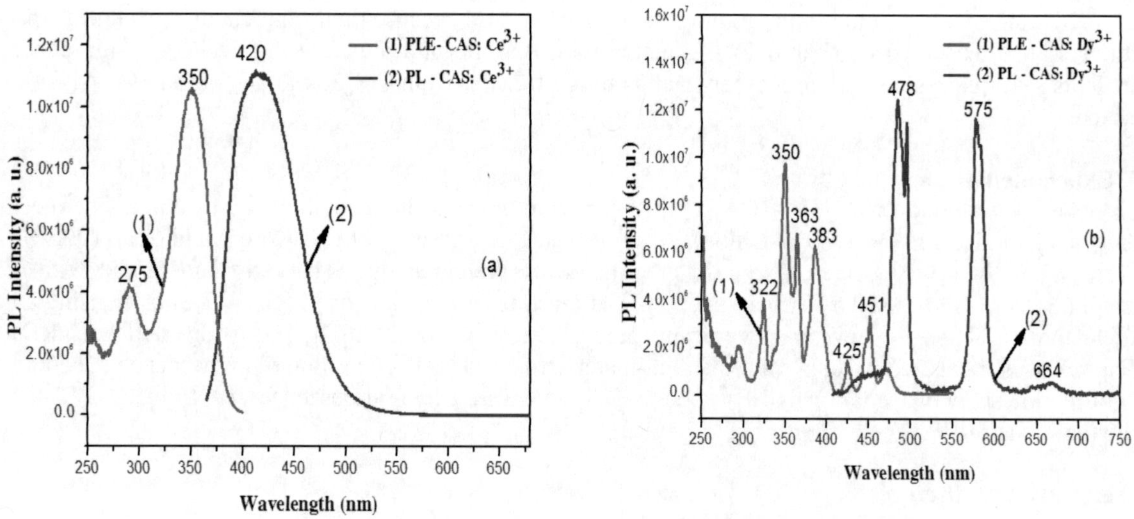

Figure 2. (a) PLE spectra of CAS: Ce^{3+} (0.5 mol%) with λ_{em}=420 nm and PL spectra of CAS: Ce^{3+} (0.5 mol%) with λ_{ex}=350 nm, (b) PLE spectra of CAS: Dy^{3+} (1.0 mol%) with λ_{em}=575 nm and PL spectra of CAS: Dy^{3+} (1.0 mol%) with λ_{ex}=350 nm

IOP Conf. Series: Materials Science and Engineering **540** (2019) 012001 doi:10.1088/1757-899X/540/1/012001

Figure 2(a) shows the photoluminescence excitation (PLE) and emission (PL) spectra of Ce^{3+} doped CAS. The PLE spectra of CAS: Ce^{3+} (line 1) monitored at 420 nm exhibits two distinct excitation bands at 275 and 350 nm, which is assigned to the 4f–5d transitions of Ce^{3+}. Besides, the PL spectra of Ce^{3+} ions in CAS (line 2) has a broad band ranging from 365 nm – 525 nm with a maximum at 420 nm, which is attributed to the 5d-4f transition of Ce^{3+}[6-8], which has high intensity. Meanwhile, Figure 2(b) the PLE spectra of CAS: Dy^{3+} (line 1) monitored at 575 nm, shows characteristic narrow lines corresponding to f-f absorption transition of Dy^{3+}[9], which are appeared at 322 nm ($^6H_{15}\rightarrow{}^6P_{3/2}$), 350 nm ($^6H_{15}\rightarrow{}^4M_{15/2},{}^6P_{7/2}$), 363 nm ($^6H_{15}\rightarrow{}^4I_{11/2}$), 383 nm ($^6H_{15}\rightarrow{}^4I_{13/2}$, $^4F_{7/2}$), 425 nm ($^6H_{15}\rightarrow{}^4G_{11/2}$,), 451 nm ($^6H_{15}\rightarrow{}^4I_{15/2}$). The emission spectra of CAS: Dy^{3+} show narrow lines at wavelength 478 nm, 575 nm, 664 nm, correspond to the transitions $^4F_{9/2}\rightarrow{}^6H_J$ (J = 15/2, 13/2, 11/2) of Dy^{3+} ion. However, their emission intensity is weaker than that of CAS: Ce^{3+}. PLE spectra of CAS: Dy^{3+} (1.0 mol%) and PL spectra of CAS: Ce^{3+} (0.5 mol%) showed that there is an overlap between the emission band of Ce^{3+} ion and the excitation spectra of Dy^{3+} ion. It could indicate the energy transfer from Ce^{3+} ion to Dy^{3+} ion in the lattice CAS.

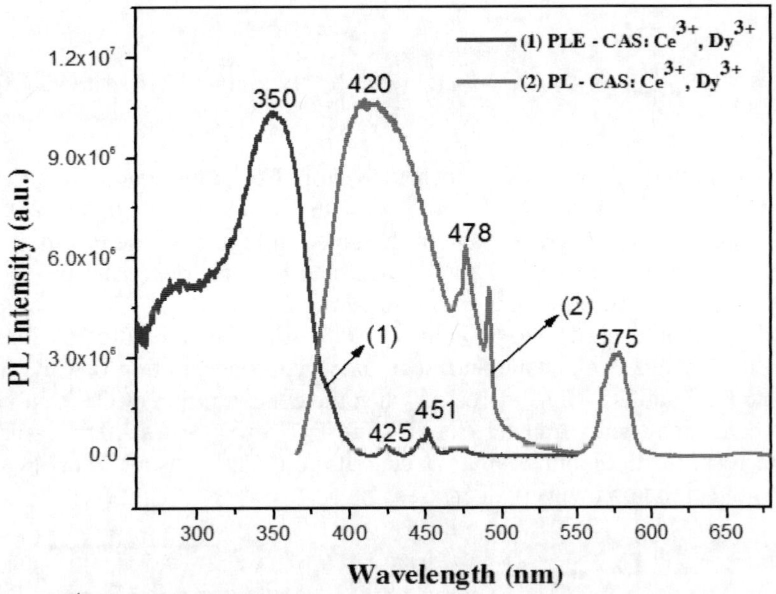

Figure 3. PLE spectra of CAS: Ce^{3+} (0.5 mol%), Dy^{3+} (1 mol%) (λ_{em} = 575 nm) and PL spectra of CAS: Ce^{3+} (0.5 mol%), Dy^{3+} (1 mol%), λ_{ex}=350 nm

Figure 3 show the PLE spectra of CAS: Ce^{3+}, Dy^{3+} (line 1) monitored at 575 nm ($^4F_{9/2}\rightarrow$ $^6H_{13/2}$ emission Dy^{3+}), consists of the excitation bands of Ce^{3+} and Dy^{3+} ions, indicating that beside to f-f absorption transition of Dy^{3+} ion[9], the absorption of Ce^{3+} ion also contributes greatly to the emission of Dy^{3+} ion. The contribution to the direct excitation of Dy^{3+} ion is negligible, mainly due to the absorption of Ce^{3+} to generate emission of Dy^{3+} ion via energy transfer from Ce^{3+} ion to Dy^{3+} ion. The result proves that the emission efficiency of Dy^{3+} ion is rather high in the CAS co-doped with Ce^{3+} and Dy^{3+} ions.

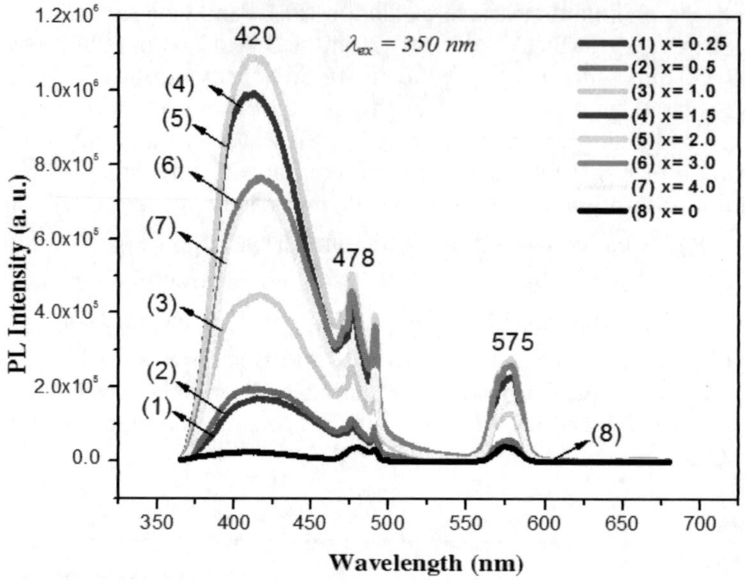

Figure 4. PL Spectra of CAS co-doped with Ce^{3+} (x mol%), Dy^{3+} (1 mol%), λ_{ex} = 350 nm

To better understand the energy transfer mechanism from Ce^{3+} ion to Dy^{3+} ion in CAS: Ce^{3+}, Dy^{3+}. The PL spectra of CAS: Ce^{3+}(x mol%), Dy^{3+} (1 mol%) with x = 0.25, 0.5, 1.0, 1.5, 2.0, 3.0, 4.0 mol% and the concentration of Dy^{3+} ion 1 mol% have investigated in Figure 4. The results show that when x = 0 mol% the emission intensity of CAS: Dy^{3+} is weak and only characteristic peaks of Dy^{3+} ions in the host could be observed. The broad band with maximum emission at 420 nm was attributed by Ce^{3+} ions, narow lines ranging from 450 to 600 nm, which was attributed by the transitions of Dy^{3+} ion. The PL spectra do not change peak shape and position when concentration of Ce^{3+} ions varies. The PL intensity of both Ce^{3+} and Dy^{3+} ions increases when the concentration of Ce^{3+} ion increases. The result confirms that there is an energy transfer from Ce^{3+} to Dy^{3+} in CAS: Ce^{3+}, Dy^{3+}. The emission intensity of Dy^{3+} ions increases with the increasing concentration of Ce^{3+} ions and reaches a maximum when x = 2 mol%, later the emission intensity decreases.

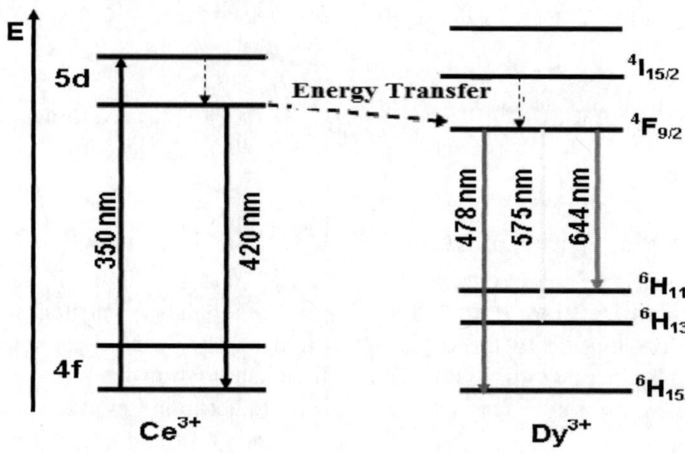

Figure 5. Energy transfer mechanism from Ce^{3+} ion to Dy^{3+} ion in CAS.

IOP Conf. Series: Materials Science and Engineering **540** (2019) 012001 doi:10.1088/1757-899X/540/1/012001

The energy level diagram of Ce^{3+}, Dy^{3+} ions and The energy transfer Ce^{3+} to Dy^{3+} in CAS: Ce^{3+}, Dy^{3+} phosphor upon excitation radiation λ_{ex}=350 nm is illustrated in Figure 5. The electrons are excited from the ground state 4f to the excited state 5d of Ce^{3+} ion and non-radiatively relaxed to the lowest state of the 5d level. Subsequently, the transition of electron from 5d state to 4f state formed broad band emission of Ce^{3+}ion. At the same time, amount electrons transferred from 5d state of Ce^{3+} ion to $^4F_{9/2}$ excitation state of Dy^{3+} ion, later transited to ^6H_J ground state of Dy^{3+} ion to generate the emission.

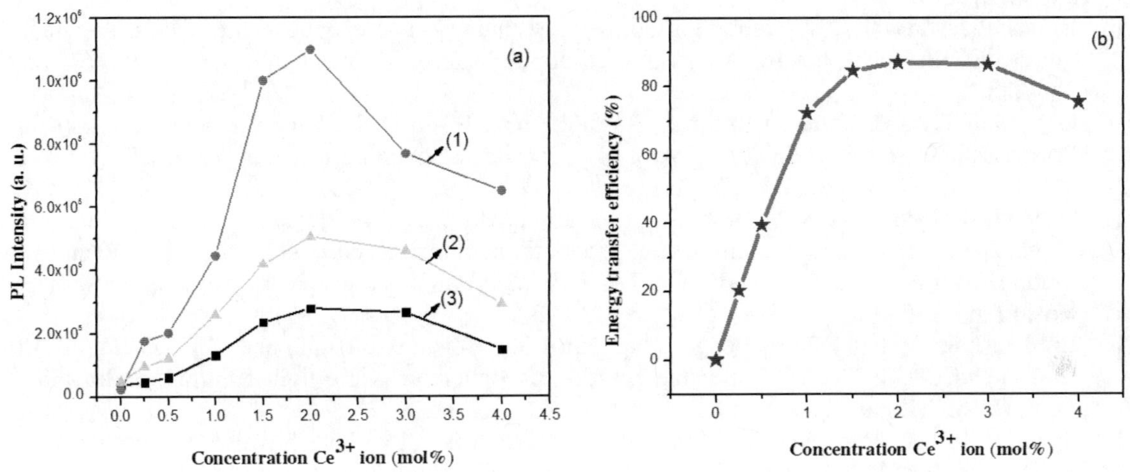

Figure 6. The dependence of PL intensity (peak at 420 (1), 478 (2) and 575 nm (3)) (a) and the energy transfer efficiency from Ce^{3+} to Dy^{3+} (b) on the concentration of Ce^{3+} ion

Figure 6 (a) shows the variation of the emission intensities of Ce^{3+} ion and Dy^{3+} ion versus the Ce^{3+} concentration. The observed variations of the emission intensity of the Ce^{3+} and Dy^{3+} ions indicate that the energy transfer process occurred from Ce^{3+} to Dy^{3+}. Thus, the energy transfer efficiency from Ce^{3+} ion to Dy^{3+} ion can be calculated by using the following formula[7]:

$$\eta = 1 - \frac{I_{S0}}{I_S}$$

Where, η is the energy transfer efficiency, I_{S0} is the intrinsic luminescence intensity of an activator (here is Dy^{3+}) in the absence of a sensitizer, I_s the luminescence intensity of Dy^{3+} in the presence of a sensitizer (here is Ce^{3+}). From Figure 6 we can see that the luminescence intensity of Dy^{3+} ion reaches a maximum when the concentration of Ce^{3+} ion is about 2 mol%. The calculation result shows that the sample CAS: Ce^{3+} (2 mol%), Dy^{3+} (1 mol%) has the highest energy transfer efficiency value of 87% as show in Figure 6(b).

4. Conclusion

Ce^{3+} and Dy^{3+} ions doped $Ca_2Al_2SiO_7$ phosphors were prepared by the solid-state reaction. Photoluminescence of CAS: Ce^{3+} shows an asymmetric broad band peaking at 420 nm, due to the transition from the 5d excited state to the $^2F_{5/2}$ and $^2F_{7/2}$ ground states. The PL spectra of CAS: Dy^{3+} shows narrow lines at wavelength 478 nm, 575 nm, 664nm correspond to the state transitions $^4F_{9/2} \rightarrow {}^6H_J$ (J = 15/2, 13/2, 11/2). The energy transfer from Ce^{3+} ion to Dy^{3+} ion appears in the Ce^{3+} ion and Dy^{3+} ion co-doped phosphor CAS: Ce^{3+}, Dy^{3+}. The highest energy transfer efficiency is optimal when doping concentration is about 2.0 mol% Ce^{3+} and 1.0 mol% Dy^{3+}.

IOP Conf. Series: Materials Science and Engineering **540** (2019) 012001 doi:10.1088/1757-899X/540/1/012001

References

[1] Koen Van den Eeckhout, Dirk Poelman and Philippe F. Smet 2013 Persistent Luminescence in Non-Eu^{2+}-Doped Compounds. *Materials*, 6, pp. 2789-2818.

[2] Nobuhiro Kodama et al 1999 Long-lasting phosphorescence in Ce^{3+} -doped Ca2Al2SiO7 and CaYAl3O7 crystals. *Applied Physics Letters*, Vol. 75, No. 12, pp. 1715-1717.

[3] Geetanjali Tiwari et al 2016 Enhanced long-persistence of $Ca_2Al_2SiO_7$:Ce^{3+} phosphors for mechanoluminescence and thermoluminescence dosimetry.*J. Mater Sci: Mater Electron*, Vol. 27, pp. 6399-6407.

[4] Ishwar Prasad Sahu 2015 Retraction notice to "Luminescence enhancement of bluish – green $Sr_2Al_2SiO_7$:Eu^{2+} phosphor by dysprosium co-doping". *Journal of Luminescence*, Vol. 167, pp. 278-288.

[5] Geetanjali Tiwari., Nameeta Brahme., Ravi Sharma., Bisen D. P., Sanjay Kumar Sao., Ishwar Prasad Sahu 2016 $Ca_2Al_2SiO_7$: Ce^{3+} phosphors for mechanoluminescence dosimetry.*The journal of biological and chemiscal luminescence*, Vol. 23, No. 1, pp. 15-23.

[6] Geetanjali Tiwari., Nameeta Brahme., R. Sharma., Bisen D. P., Sanjay Kumar Sao., Manisha Singh 2015 Fracto-mechanoluminescence and thermoluminescence properties of UV and γ-irradiated $Ca_2Al_2SiO_7$: Ce^{3+} phosphor. *The journal of biological and chemiscal luminescence,* Vol. 12, pp. 155-163.

[7] Li Pan-Lai, Wang Yong-Sheng, Zhao Su-Ling, Zhang Fu-Jun and Xu Zheng 2012 Ca_2BO_3Cl:Ce^{3+},Tb^{3+}: A novel tunable emitting phosphor for white light-emitting diode. *Chin. Phys. B* Vol. 21, No. 12.

[8] Sandeep Kumar., Ram Prakash., Vinay Kumar 2015 A novel yellowish white Dy^{3+} activated α-Al_2O_3 phosphor: Photoluminescence and optical studies. *Functional Materials Letters,* Vol. 8, No. 6, pp. 1550061(4 pages).

[9] Nita Shinde, N. S. Dhoble, S. C. Gedamc and S. J. Dhoblea. (2014). Photoluminescence enhancement in Na_3SO_4Cl: X (X = Ce^{3+}, Eu^{3+} or Dy^{3+}) material. The journal of biological and chemical luminescence, Vol. 30, pp. 898-903

IOP Conf. Series: Materials Science and Engineering **540** (2019) 012002 doi:10.1088/1757-899X/540/1/012002

Effect of Organic Plating Additives Concentration on Carbon Content and Donor Density of Electroplated Tin

Marie Angelynne Fabro[1] and Manolo Mena[2]

1 Instructor, Department of Mining, Metallurgical and Materials Engineering, University of the Philippines, Quezon City 1101, Philippines
2 Professor, Department of Mining, Metallurgical and Materials Engineering, University of the Philippines, Quezon City 1101, Philippines
E-mail: mcfabro@up.edu.ph

Abstract. Pure tin is widely used as a replacement of Pb-Sn eutectic alloy as finish and solder material. This study aims to determine how the increasing concentration of organic plating additives affects the carbon content and donor density of electroplated Sn. Pure tin was electroplated on an Al sheet in a Hull cell using a commercial plating solution. Plating was performed in electrolyte baths with primary additive concentration from 100 mL/L to 600 mL/L, and with secondary additive concentration from 8 mL/L to 48 mL/L. Current densities of 15, 8 and 3 A/dm^2 were considered. The increased concentration of organic plating additives caused an increase in carbon content of the deposit from 5 wt.% C to 8 wt.%C. Mott-Schottky measurements showed increasing electron donor density of up to four orders of magnitude as a function of current density and carbon content. A higher current density during deposition causes the tin oxide film to reach the donor density saturation value at lower concentrations of organic plating additives. This suggests that current plays a role in the decomposition of organic additives, which results in the incorporation of carbon in the deposit.

1. Introduction

The search for a Pb-free alternative to the Pb-Sn eutectic alloy as solder material and surface finish in microelectronics applications has prompted the investigation of a wide range of alloys over the last three decades. This search is driven by the need for compliance with global and regional environmental regulations. [1] While pure Sn is an obvious alternative, whisker growth, void formation and corrosion of Sn finish under high temperature and high humidity conditions are among the reliability issues that the industry must address. [2]

After the application of the pure Sn layer, a thin and stable native oxide rapidly forms, which protects the Sn from further oxidation. Sometimes, however, the oxidation of Sn continues when exposed to high temperatures such as reflow processes. This runaway oxidation results to a yellow, blue, purple or brown color. [3,4] The discoloration becomes an indicator of poor wettability and solderability. [4]

This study aims to determine the effect of organic additives of a Sn electroplating bath and current density on the carbon content and donor density of the electroplated Sn.

The Mott-Schottky equation allows for the calculation of the donor density of a semiconductor. Changing the voltage of the semiconductor artificially, while it is in contact with an electrolyte, causes a change in the net charge in the semiconductor, the opposing charge formed by the electrolyte, and the thickness of the space charge region. This results in a change in capacitance as well. The change in the interfacial capacitance can be related to the applied bias potential using the Mott-Schottky equation:

$$\frac{1}{C^2}=\frac{2}{{}_{o}A^2 \mathrm{e} N_\mathrm{D}}\left(V \quad V_\mathrm{fb} \quad \frac{\mathrm{k_B}T}{\mathrm{e}}\right) \tag{1}$$

where C is the interfacial capacitance, A is the area, N_D is the number of donors per unit volume, V is the applied voltage, k_B is Boltzmann's constant, T is absolute temperature, and e is the electronic charge. [5]

2. Methodology

2.1. Electroplating of pure Sn

Pure Sn was electroplated using a commercially available Solderon™ matte Sn electrolyte in a Hull cell. The composition of the plating solution was varied as shown in table 1. Pure Sn was used as the anode, while the cathode substrate was an aluminum plate. Prior to electroplating, the substrate was wiped clean with alcohol and washed with deionized water, and dipped in 50 g/L KOH for 2 minutes to activate the surface. It was given a final wash with deionized water prior to the start of electroplating. Electroplating was performed using a current of 3.0 A, at an average plating bath temperature of 47°C.

The plating bath has five components: Sn concentrate, acid, primary additive, secondary additive and RD concentrate. The primary additive is required to achieve smooth, uniform deposits. On the other hand the secondary additive is added to achieve uniform deposit thickness. The proprietary nature of the plating bath allows only admission that the primary and secondary additives are composed primarily of organic compounds. Lastly, the RD concentrate is added to minimize the oxidation of stannous tin. [6]

Table 1. Electroplating bath compositions with varying plating additives concentrations

Organics Level	1	2	3	4	5	6
Tin	80 g/L					
Acid	210 mL/L					
Primary (mL/L)	100	200	300	400	500	600
Secondary (mL/L)	8	16	24	32	40	48
Concentrate	18 mL/L					

The Hull cell allowed for the simultaneous deposition at a range of current densities. Table 2 shows the current densities considered in this study.

Table 2. Electroplating current densities

	Current density (A/dm^2)
High	15
Medium	8
Low	3

2.2. Carbon content determination

Samples for carbon content determination were prepared by cutting a 5mm by 5mm piece of electroplated Sn at the location corresponding to the desired current density. The samples were inspected using a JEOL JSM-6010LV Scanning Electron Microscope with Energy Dispersive X-Ray Analysis (SEM-EDX). Analysis was performed on three points of each sample, at accelerating voltages of 5 kV.

2.3. Donor density measurement

Donor density determinations using Mott-Schottky plots were conducted using a three-electrode setup attached to a potentionstat (Metrohm Autolab PGSTAT128N) controlled using the NOVA 2 software. The as-plated Sn samples were used as working electrodes (WE). A platinum electrode was used as the counter electrode (CE), while a Ag|AgCl equilibrated in saturated KCl solution was used as the reference electrode (RE). The working electrolyte was 3.5% NaCl solution that has been purged with pure nitrogen gas prior to electrochemical measurements. A Luggin capillary with diameter 1 mm was used as an electrolyte bridge between the WE and the RE. All experiments were done at room temperature.

Mott-Schottky plots for as-plated samples were generated from EIS measurements made from 100 kHz to 10 kHz, from 0.2V to 0.8V bias potentials at increments of 0.05V. At these frequencies, the measurements were determined to be in the charge transfer controlled regime.

3. Results and discussion

Figure 1 below shows the carbon content of as-plated Sn at the range of organic levels and three current densities, measured by EDX at 5kV. The carbon content of Sn deposited at low current density increases from about 5 wt.% at the lowest organics level setting to about 7 wt.% at the highest organics setting. On the other hand the carbon content of Sn deposited at medium current density increases from about 5 wt.% to 8 wt.% over the same range. The carbon content of Sn electroplated at high current density, however, shows no trend. This erratic behavior may be attributed to the roughness of the sample.

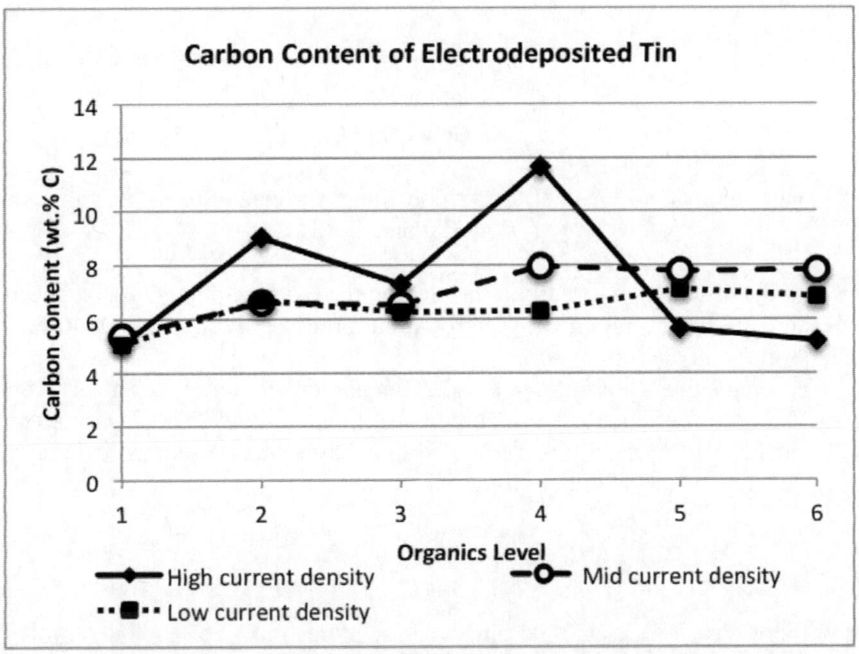

Figure 1. Carbon content of as-plated Sn, measured at 5kV.

The carbon content of 5 wt.% at the lowest organic level setting is about two orders of magnitude higher than the co-deposited carbon content that is quoted by the plating bath manufacturer for comparable bath compositions. If this is accurate, then the presence of excess carbon must be attributed to some other mechanism.

One study, upon observation of significant porosity in their electroplated Sn, suggested that the organic component of the plating bath may become trapped in the voids. They also point to this possibility as the cause of the eventual corrosion of reflowed Sn. [4]

The sulfonate-based Solderon plating bath may possibly degrade at a significant rate. The degradation of the sulfopropyl sulfonate additive in acid copper electroplating baths has been studied. Sulfopropyl sulfonate degrades in un-deaerated conditions, while the presence of the Cu(I) ion increases the decomposition rate. Even if the nominal concentrations of the additives are maintained, the presence of decomposition products of the additives affects the super filling capability of the bath. This may result in the presence of voids in the deposit, where organic molecules may become trapped. [7,8]

The depth profile of as-plated Sn in figure 2, taken using Time-of-Flight Secondary Ion Mass Spectrometry, supports this hypothesis.

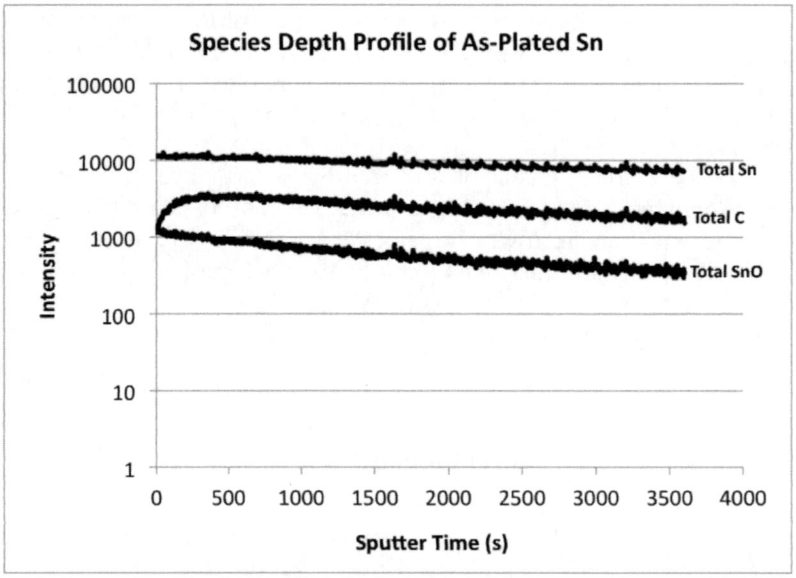

Figure 2. Secondary ion mass spectrometry depth profile of as-plated Sn, with estimated crater depth of 350 nm.

Figure 2 shows that the distribution of carbon throughout the oxide and the metal suggests that carbon is included into the deposit throughout the electroplating process, and that the inclusion of carbon is not an occasional event.

Mott-Schottky measurements, which allow for the calculation of donor densities, were made only from 100kHz to 10kHz to minimize the measurement time and avoid unnecessary alterations on the sample. These frequencies were determined to be within charge transfer-controlled conditions. Figure 3 below shows an example Mott-Schottky plot.

IOP Conf. Series: Materials Science and Engineering **540** (2019) 012002 doi:10.1088/1757-899X/540/1/012002

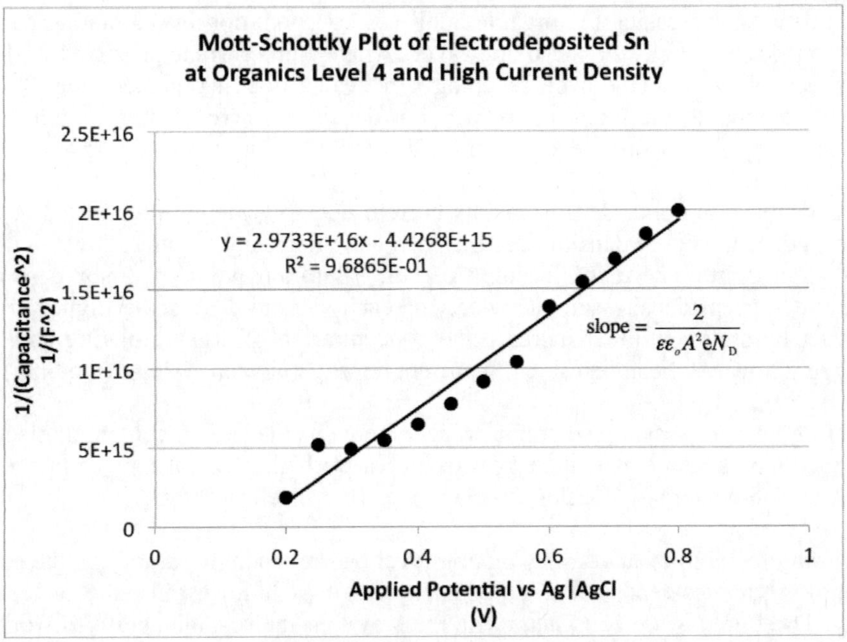

Figure 3. Mott-Schottky Plot of as-plated Sn at organics level 4 and high current density from EIS measurements made at 100kHz.

The linear and sloping regions of the Mott-Schottky plots all have positive slopes, which is characteristic of n-type semiconductors. [5] Mott-Schottky plots could have allowed for the calculation of the donor density. However the composition of the oxide layer is not exactly known; it may be an intimate mixture of SnO and SnO_2, an outer film of SnO_2 covering SnO. [9,10] In this case the application of a dielectric constant is suspect. Thus figure 4 below shows comparative values of donor density across settings.

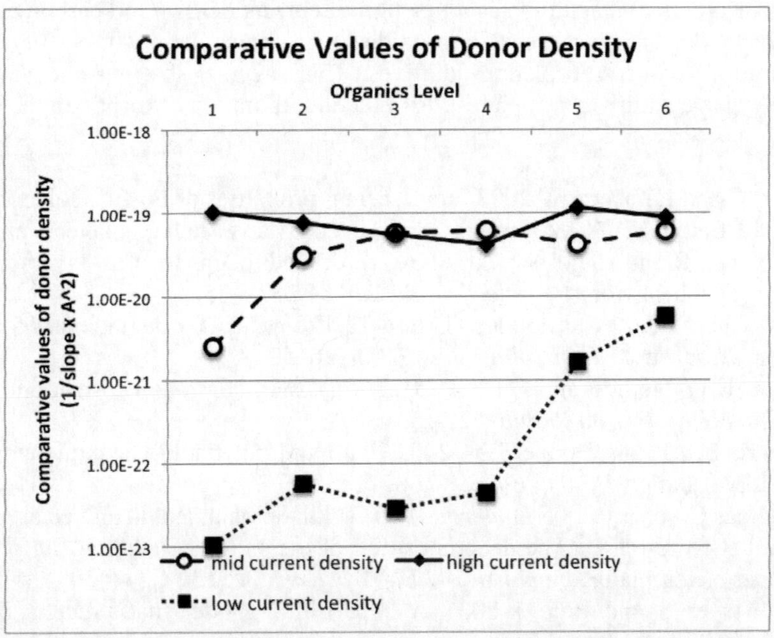

Figure 4. Comparative values of donor density of as-plated Sn measured at 100 kHz.

At high current density, increasing the organic additives concentration in the plating bath appears to have no effect on donor density of the oxide layer. At medium current density, the donor density increases with increasing organics level, reaching a plateau value at organics level 3. This plateau value appears to be equal to the single value of donor density observed at high current density. This suggests that increasing the organics level increases the donor density in the native oxide layer, up to a certain saturation point.

The continuous increase in donor density of low current density samples supports the existence of a saturation point. While the donor density does increase with increasing organics level, it does not quite reach the saturation point. The existence of a certain saturation point of donor density would be definite if the low current density samples were to reach saturation at some higher organics level. However it is not possible to further increase the concentration of organic plating additives beyond level 6. This might however be observable if a current density between $8A/dm^2$ (medium) and $3A/dm^2$ is selected.

As discussed in figure 1, increasing the organics level increases the carbon content of electroplated Sn. The increase in donor density may then be associated with the increase in carbon content. This suggests that a carbon-containing species, or perhaps carbon itself, acts as a dopant in the native tin oxide film.

Another noteworthy feature is that at every plating level before reaching saturation, the electron donor density is highest in samples deposited at high current density, and lowest in samples deposited at low current density. These trends are consistent with the way that the organic additives work to improve the characteristics of the plated deposit. Organic additives that are intended to help level the deposit should be adsorbed at regions of high current density. This corresponds to bumps or regions that are higher or thicker than the surrounding deposit. [8,11,12] Levelers, which have amine functional groups or aromatic rings, adsorb at regions of high current density. [13] This suggests that current plays a role in the decomposition of organic plating additives that results in the incorporation of carbon in the deposit.

4. Conclusions

The increased concentration of organic plating additives caused an increase in carbon content of the deposit from 5 wt.% C to 8 wt.%C. Mott-Schottky measurements showed increasing electron donor density of up to four orders of magnitude as a function of current density and carbon content. A higher current density during deposition causes the tin oxide film to reach the donor density saturation value at lower concentrations of organic plating additives. This suggests that current plays a role in the decomposition of organic additives, which results in the incorporation of carbon in the deposit.

5. References

[1] Handwerker C and Lafayette W 2007 Fundamental properties of Pb-free solder alloys *Lead-free soldering* ed J Bath (New York: Springer Science+Business Media) chapter 2 pp 21-74

[2] Eckold P, Niewa R and Hügel W 2014 Texture of electrodeposited tin layers and its influence on their corrosion behavior *Microelectron Reliab.* **54** 2578–85

[3] Chung C K, Chen Y J, Li C C and Kao C R 2012 The critical oxide thickness for Pb-free reflow soldering on Cu substrate *Thin Solid Films* **520** 5346–52

[4] Zeng X, Sun H Q, He Y F and Qu X P 2012 Reflow discoloration formation on pure tin (Sn) surface finish *Microelectron Reliab.* **52** 1153–6

[5] Gelderman K, Lee L and Donne S W 2007 Flat-band potential of a semiconductor: using the Mott–Schottky equation *J Chem Educ.* **84** 685

[6] Rohm and Haas Company 2005 *Solderon* ^{TM} *SC* (Philadelphia: Rohm and Haas Company)

[7] Frank A and Bard A J 2003 The decomposition of the sulfonate additive sulfopropyl sulfonate in acid copper electroplating chemistries *J Electrochem Soc.* **150** C244–50

[8] West AC, Mayer S and Reid J 2001 A superfilling model that predicts bump formation *Electrochem Solid-State Lett.* **4** C50

[9] Morgan Tench D, Anderson DP and Kim P 1994 Solderability assessment via sequential electrochemical reduction analysis *J Appl Electrochem.* **24** 18–29

IOP Conf. Series: Materials Science and Engineering **540** (2019) 012002 doi:10.1088/1757-899X/540/1/012002

[10] Cho S, Yu J, Kang S K and Shih D Y 2005 Oxidation study of pure tin and its alloys via electrochemical reduction analysis *J Electron Mater.* **34** 635–42

[11] Cao Y, Taephaisitphongse P, Chalupa R and West A C 2001 Three-additive model of superfilling of copper *J Electrochem Soc.* **148** C466

[12] Kelly J J, Tian C and West A C 1999 Leveling and microstructural effects of additives for copper electrodeposition *J Electrochem Soc.* **146** 2540–5

[13] Lee H and Chen C-M 2018 Impurity effects in electroplated-copper solder joints *Metals* **8** 388

Acknowledgements

The authors would like to thank the following for their contributions to this project: Prof. Candy C. Mercado and Prof. Alberto V. Amorsolo, Jr., Nexperia Philippines, Inc. (formerly NXP Semiconductors Philippines, Inc.), particularly Ms. Liza T. Lopez. M Fabro further acknowledges support from her family and friends.

IOP Conf. Series: Materials Science and Engineering **540** (2019) 012003 doi:10.1088/1757-899X/540/1/012003

Synthesis and Characterization of a g-C_3N_4 Coupled Hybrid Nanocomposite for the Photocatalytic Effects

D P Ojha[1], J H Song[1] and H J Kim[1,2,3]

1 Department of Convergence Technology Engineering, Chonbuk National University, Jeonju 561-756, Republic of Korea
2 Eco-friendly Machine Parts Design Center, Chonbuk National University, Jeonju 561-756, Republic of Korea
3 Email: khanjoo@jbnu.ac.kr

Abstract. Photodegradation of organic pollutants using nanoparticles with suitable band gap is one of the most studied technologies in last few decades. About 8 nm antimony-doped tin oxide (ATO) nanoparticles, as the photocatalyst for organic degradation, is prepared by the calcination of the stoichiometric mixture of precursor hydroxides of Sn^{4+} and Sn^{3+}. ATO was combined with thermally synthesized graphitic carbon nitride(g-C_3N_4 or g-CN) and the resulting Z-scheme g-CN/ATO nanocomposite was utilized for the decomposition of salicylic acid (SA) in aqueous solution. All the samples were analyzed by scanning electron microscopy (SEM), transmission electron microscopy (TEM), X-ray diffraction (XRD), energy dispersive spectroscopy (EDS), Fourier transform infrared spectroscopy (FTIR) and the progress of the photocatalytic degradation reaction was measured by using UV-Visible spectroscopy. The rate constant measurements showed that the rate of degradation of SA is enhanced with hybrid nanocomposite.

1. Introduction

Heterogeneous photocatalysis has been regarded as the sustainable renewable technology to solve the global energy and environmental crisis, in which semiconductor materials generate reactive electron-hole pair by band gap excitation in presence of sunlight [1]. TiO_2 is considered most promising photocatalyst due to its low cost, high stability and non-toxicity as well as suitably placed band position and excellent surface properties. However, practical application of TiO_2 is hindered due to its wide band gap (Eg = 3.2 eV) which can only absorb UV light ($\lambda < 380$ nm) which is merely a 4% of the whole sunlight [2-5]. Thus, quest of alternative photocatalysts that can utilize the major portion of the sunlight i.e. visible light is another major strategy in the photocatalysis. Antimony doped tin oxide (Sb.SnO_2: ATO) is one such semiconductor with excellent charge mobility and narrow band gap (E_g = 2.65 eV) that can absorb major portion of sunlight [6, 7]. However, firstly the band position of ATO is not appropriately placed with conduction band (CB, +0.95 V) and valence band (VB, +3.6 V); it cannot generate the reactive species in aqueous medium. Secondly, the recombination of electron-hole is faster than the electron transport [8].

Coupling of two semiconductors with matching band position can decrease the recombination rate due to the charge transfer and thus the photogenerated electron-hole pair becomes more available for higher catalytic efficiency [9, 10]. A notable strategy to overcome this limitation was applied by combining it with TiO_2 for the decomposition of Methylene Blue in a visible-light photocatalytic mechanism [11]. Next, graphitic carbon nitride (g-C_3N_4 or g-CN) could be a potential material for yet

IOP Conf. Series: Materials Science and Engineering **540** (2019) 012003 doi:10.1088/1757-899X/540/1/012003

another strategy. g-C_3N_4 has sufficiently negative conduction band position to utilize the photogenerated electron and its exfoliated 2D flat-structure provides more reactive sites that reduces e-h recombination rate [12]. Thus, a new z-scheme type ATO/g-C_3N_4 system is designed. The photocatalytic decomposition of a model organic compound salicylic acid (SA) was studied and the origin of the enhanced photoactivity was suggested in a mechanism.

2. Material and methods

2.1. Materials
All the reagents, melamine, tin chloride ($SnCl_4.5H_2O$), sodium bicarbonate (NH_4HCO_3), acetic acid (CH_3COOH), antimony chloride ($SbCl_3$), ethanol, salicylic acid (SA) and NH_4OH were purchased from Sigma-Aldrich and used directly for experiments without any further purification.

2.2. Synthesis of g-CN nanosheet
Melamine was thermally polymerized at 550 °C for 3 hr in a muffle furnace resulting yellowish graphitic carbon nitride (g-CN) [13]. g-CN powder was crushed into fine powder and again heated at 550 °C for 3 hr to complete the polymerization of the remaining melamine. Next, the bulk g-CN powder was stirred overnight with oxalic acid (1.0 M). Thus, exfoliated g-CN nanosheet was separated from the suspension by filtration, freeze-dried and heated at 300 °C to remove the residual oxalic acid.

2.3. Synthesis of ATO nanoparticles and ATO/g-CN nanocomposite
In the synthesis of ATO nanoparticles as reported in [14], 50 ml of ethanol was vigorously stirred with 3.25 g of $SnCl_4 \cdot 5H_2O$, 5 mol% $SbCl_3$, 5.5 g NH_4HCO_3 and 1.75 g CH_3COOH in a 100 ml beaker for 4 hr. The pH of the solution was adjusted at 7 by adding ammonia solution. The white suspension of the $Sn(OH)_4$ and $Sb(OH)_3$ was centrifuged and washed 3 times with distilled water and calcined at 550 °C for 8 h. The bluish grey powder confirmed the formation of ATO nanopowder. For the synthesis of ATO/g-CN nanocomposite, ATO NP and g-CN were separately sonicated in ethanol: water (1:1) solution for 30 min and mixed together before a hydrothermal treatment at 180 °C for 4 hr. The resulting composite with 1%, 5% and 10% by weight of g-CN was washed several times and dried.

2.4. Characterization
The crystal phase of the prepared samples was identified by X-ray powder diffraction (Rigaku Multiflex diffractometer). Likewise, the surface morphology of all samples was analyzed by a field-emission scanning electron microscope (FE-SEM, S-7400, Hitachi, Japan). Furthermore, the morphology and crystal phase were also analyzed using high-resolution transmission electron microscope (HRTEM, JEM-2010F-JEOL Japan).

2.5. Photocatalytic measurement
The photoactivity of the samples was studied in terms of the photodegradation of SA with visible light source [15]. 1 mg/l of the catalyst and 50 ml of the 10 mg/l SA aqueous solution were stirred in a 100 ml photo-reactor for 30 min in dark to attain the adsorption-desorption equilibrium between the catalyst and SA. The solution was irradiated with visible light from A 300 W Xe lamp (DY. Tech., Korea) equipped with a UV and IR cut-off filters (Edmund optics, USA). After certain time interval, 2 ml aliquot of the solution was drawn out with a syringe fitted with a filter and the absorbance was measured in a UV-Visible spectrophotometer. The concentration of the remaining SA corresponding to the the absorbance at 297 nm was calculated from the Beer-Lambert Law. Finally, the rate of the photodegradation of the various samples was compared by the following equation:

$$Degradation\ efficiency\ (\%) = 1 - \frac{C}{Co} \times 100\%$$

Here, C_0 is the concentration of SA at 0 min, and C is the concentration of residual SA at a certain time (t). The dark reaction was also performed for the comparison.

IOP Conf. Series: Materials Science and Engineering **540** (2019) 012003 doi:10.1088/1757-899X/540/1/012003

3. Results and discussion

3.1. *Morphological studies*

XRD pattern in figure 1(a) shows the diffraction peaks of SnO_2 with tetragonal rutile structure corresponding to the JCPDS card number 01-072-1147. No impurity peaks related to Sb_2O_3 or Sb_2O_5 and the bluish grey color of the powder indicated successful doping of Sb on the crystal lattice of SnO_2 and indicated the presence of the new energy band of ATO below the conduction band minima, originated from outer orbitals (5d and 6s) of Sb. Thus, the CB minima of SnO_2 is lowered and band gap of ATO as 2.55 eV [16]. XRD pattern of 10% ATO/g-CN composites revealed no characteristic diffraction g-CN in the nanocomposite likely due to the low g-CN content as shown in figure 1(b)) [17].

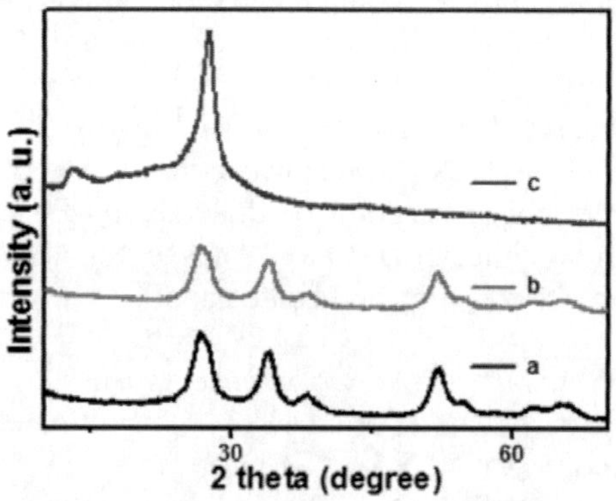

Figure 1. XRD spectra of (a) ATO nanoparticles, (b) ^{10}ATO/g-CN nanocomposite and (c) g-CN nanosheet.

The shape size and the crystalline nature of the as-prepared nanoparticle and nanocomposites were investigated in detail by FESEM and TEM/HRTEM as shown in figure 2. Figure 2(a) and 2(b) showed FESEM images of monodispersed ATO with high aggregation and ^{5}ATO/g-CN nanocomposite respectively. The image of ^{5}ATO/g-CN nanocomposite shows the good attachment of ATO nanoparticles and the g-CN sheet. The intercalation of ATO NPs on g-CN sheet resulted in less agglomeration in the nanocomposite which would significantly increase the surface area and surface characteristics like porosity of the nanocomposite compared to individual components. A TEM image in figure 2(c) shows of clusters of uniform 8 nm ATO nanoparticles. Figure 2(d) shows the TEM image of exfoliated g-CN sheet. Figure 2(e) shows the TEM image of the ^{5}ATO/g-CN nanocomposite [18]. HRTEM of ^{5}ATO/g-CN nanocomposite shows well-developed lattice fringes of ATO nanoparticles which underlines high crystallinity in figure 2(f). The d-spacing of 0.335 nm corresponds to the 101 planes of the tetragonal rutile crystal (JCPDS: #01-072-1147).

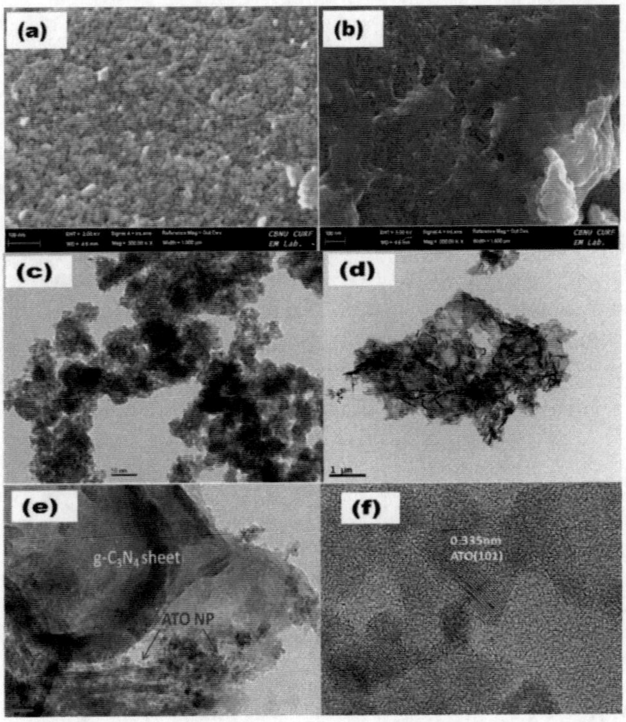

Figure 2. SEM images of (a) ATO nanoparticles and (b) ^{5}ATO/g-CN nanocomposite; TEM images of (c) ATO nanoparticles, (d) g-CN nanosheet, (e) ^{5}ATO/g-CN nanocomposite and (f) HRTEM image of ^{5}ATO/g-CN nanocomposite.

3.2. *Photocatalytic activity*

The photodecomposition of SA with different samples was assessed under visible-light irradiation. The concentration of the remnant SA was quantified from the absorption peak at 297 nm in the UV-Visible spectra with respect to different samples. Then, the concentration of solution with several photocatalytic samples was plotted as a function of photoreaction time as displayed in figure 3.

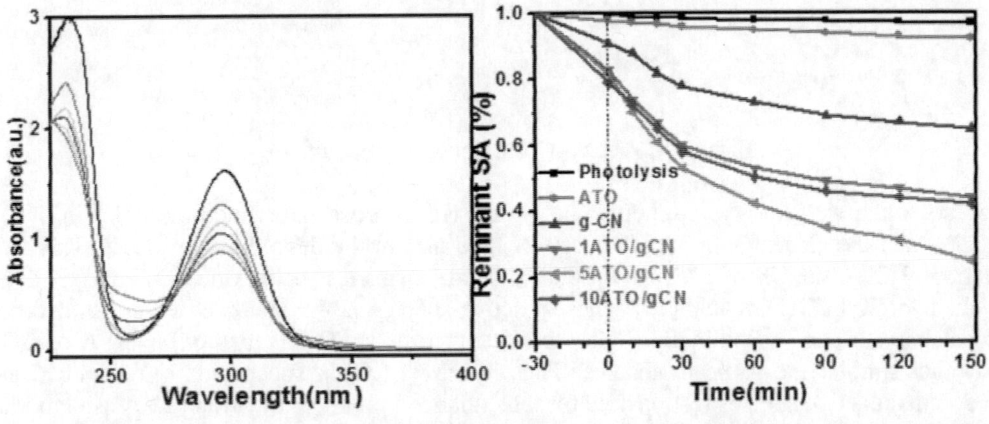

Figure 3. The absorption spectra of SA degradation over ^{5}ATO/g-CN, and the Photoactivity of different samples.

ATO/g-CN nanocomposites demonstrated higher photoactivity than the bare ATO or g-CN. Approximately 75% of SA was removed in the 1.5 hr irradiation time, whereas the bare ATO or g-CN could only remove 9% and 38% respectively. The photoactivity of ATO/g-CN nanocomposites is increased by increasing the g-CN content from 1 to 5 wt% by weight however but it is decreased above 5 wt% as in figure 3b. In brief, approximately 48%, 75%, and 54% of the SA is mineralized by 1 wt%, 5 wt% and 10 wt% ATO/g-CN nanocomposites respectively in 1.5 hr photoreaction time.
Based on the above results, the photocatalytic mechanism of [5]ATO/g-CN nanocomposite was proposed in figure 4. The narrow band gap ATO nanoparticles (2.65eV) and g-CN (2.6eV) can both absorb visible light leading to the formation of e-h pairs. Since the CB minimum of the g-CN (-1.13V) is quite negative compared to that of ATO (+0.95 V), the photogenerated electrons from the CB of g-CN to the CB of ATO cannot jump due to the steep downhill in the energy level, but CB electron of ATO easily jump to the VB of g-CN (+1.47 V). In this way, the recombination rate in both semiconductors is significantly decreased.

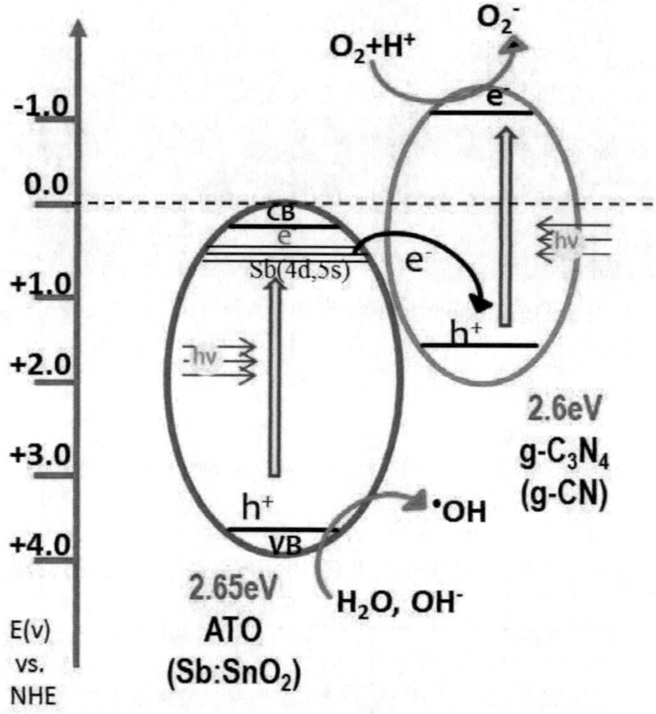

Figure 4. ATO/g-CN Z-scheme mechanism.

This suggests to the well-known Z-scheme photodegradation process in which the holes in the VB of ATO (+2.9V) and the electrons in the CB of g-CN take part in the degradation process. This leads to the reduction of O_2 at the CB of g-CN to produce the active species superoxide ($^{\bullet}O_2^-$) and oxidation of water to the hydroxyl ($^{\bullet}OH$) radical [19]. These reactive species subsequently oxidize the adsorbed SA on photocatalyst surface. Additionally, holes generated in the ATO VB also oxidize SA to CO_2 and H_2O and other simpler inorganic molecules. Thus hydroxyl ($^{\bullet}OH$), superoxide ($^{\bullet}O_2^-$) and h^+ are the three predominant active species responsible for the enhanced photodegradation of SA as elaborated in the following equations:

Photo-excitation:

$$\text{Photocatalyst} + h\nu \rightarrow (e^- + h^+)$$

Photo-reduction:

$O_2 + e^- \rightarrow {}^{\bullet}O_2^-,$

${}^{\bullet}O_2^- + e^- + 2H^+ \rightarrow H_2O_2,$

$H_2O_2 + e^- \rightarrow {}^{\bullet}OH + OH^-$

Photo-oxidation:

$2h^+ + H_2O \rightarrow 2H^+ + {}^{\bullet}OH$

Photo-degradation:

${}^{\bullet}O_2^-, {}^{\bullet}OH + SA \rightarrow$ Degradation products.

Based on above experiments the photocatalytic scheme has been designed as shown in figure 4.

4. Conclusion

Thus, monodisperse 8 nm bluish ATO nanoparticles were synthesized and applied with g-CN for the photocatalysis. Z-scheme ATO/g-CN nanocomposite showed the higher photoactivity compared to the ATO nanoparticles towards salicylic acid mineralization. The higher photocatalytic activity of ATO/g-CN over g-CN and ATO under visible light irradiation is attributed to the well-aligned band position and increased surface properties. This ultimately accelerated charge-carrier transfer rate, lowered recombination rate and subsequently enhanced photoactivity.

5. Acknowledgement

This work is supported by a grant from the Korean Ministry of Education, Science, and Technology (MEST) through the National Research Foundation (NRF) (Project No. 2017-R1C1B2011968).

6. References

[1] Tang W Z, Zhang Z, An H, Quintana M O and Torres D F 1997 *Environmental Technology* **18** 1-12
[2] Kudo A and Miseki Y 2009 *Chemical Society Reviews* **38** 253-78
[3] Kazuhito H, Hiroshi I and Akira F 2005 *Japanese Journal of Applied Physics* **44** 8269
[4] Pichat P 2016 An Overview of the Potential Applications of TiO_2 Photocatalysis for Food Packaging, Medical Implants, and Chemical Compound Delivery *Photocatalysis: Applications* ed D D Dionysiou, G L Puma, J Ye, J Schneider and D Bahnemann (Cambridge:RSC Publishing) chapter 12 pp 345-367
[5] Anpo M and Takeuchi M 2014 *Journal of Catalysis* **216** 505-16
[6] Sun Y, Chemelewski W D, Berglund S P, Li C, He H, Shi G and Mullins C B 2014 *ACS Applied Materials & Interfaces* **6** 5494-9
[7] Zhang J and Gao L 2004 *Inorganic Chemistry Communications* **7** 91-3
[8] Ojha D P, Karki H P and Kim H J 2018 *Journal of Industrial and Engineering Chemistry* **61** 87-96
[9] Schneider J, Matsuoka M, Takeuchi M, Zhang J, Horiuchi Y, Anpo M and Bahnemann D W 2014 *Chemical Reviews* **114** 9919-86
[10] Yuan L, Han C, Pagliaro M and Y.-J. Xu Y -J 2015 *The Journal of Physical Chemistry C* **120** 265-73
[11] Vilar V J P, Amorim C C, Puma G L, Malato S and Dionysiou D D 2017 *Chemical Engineering Journal* **310(2)** 329-30
[12] Cao S, Low J, Yu J and Jaroniec M 2015 *Advanced Materials* **27** 2150-76
[13] Mao J, Zhang L, Wang H, Zhang Q, Zhang W and Li P 2018 *Chemical Engineering Journal* **342** 30-42
[14] Bai F, He Y, He P, Tang Y and Jia Z 2006 *Materials Letters* **60** 3126-9
[15] Ojha D P, Joshi M K and Kim H J 2017 *Ceramics International* **43** 1290-7
[16] Rana M P S, Singh F, Joshi K, Negi S and Ramola R C 2016 *Thin Solid Films* **616** 34-42

IOP Conf. Series: Materials Science and Engineering **540** (2019) 012003 doi:10.1088/1757-899X/540/1/012003

[17] Muñoz-Batista M J, Gómez-Cerezo M N, Kubacka A, Tudela D and Fernández-García M 2014 *ACS Catalysis* **4** 63-72
[18] Chen Y, Huang W, He D, Situ Y and Huang H 2014 *ACS applied materials & interfaces* **6** 14405-14
[19] Zhang J and Nosaka Y 2014 *The Journal of Physical Chemistry C* **118** 10824-32

IOP Conf. Series: Materials Science and Engineering **540** (2019) 012004 doi:10.1088/1757-899X/540/1/012004

Adsorption of Cu(II) onto Fe(III)-Modified Montmorillonite – Kinetic, Isotherm, and Thermodynamic Studies

John Kenneth Cruz[1] and Leslie Joy Diaz[1,2]

1 Department of Mining, Metallurgical and Materials Engineering, University of the Philippines Diliman, Quezon City 1101, Philippines
2 Environmental Engineering Program, University of the Philippines Diliman, Quezon City 1101, Philippines
E-mail: jacruz1@up.edu.ph

Abstract. Trace metal ions, such as Cu(II), in wastewater could pose serious health threats when not treated. This study employed adsorption using an Fe(III)-modified montmorillonite (Fe-MMT), an abundant mineral in the Philippines, to remove Cu(II) ions in synthetic wastewater. Scanning electron micrographs of the synthesized clay show showed its layered structure while elemental analysis confirmed the successful modification to Fe-MMT with high amounts of retained Fe in the clay. Batch adsorption was performed at varying adsorption time, initial Cu(II) concentration, and temperature, for which applicable models were then employed to understand the adsorption mechanism. Cu(II) adsorption showed a slow rate until the process reached equilibrium at 12 hrs corresponding to 85.86% removal. Adsorption of Cu(II) on Fe-MMT was found to agree with both the pseudo-first order and the liquid-film diffusion kinetic models. Isotherm modelling revealed that the process fitted well with the Langmuir model with theoretical adsorption capacity of Fe-MMT as 48.077 mg/g. On the other hand, Dubinin–Radushkevich isotherm suggests that the mechanism was via physisorption. From the thermodynamic modelling, it was obtained that the Cu(II) adsorption was an endothermic process which further translated to a ΔG°_{298} of 5,696.89 J/mol.

1. Introduction

Untreated dissolved metals in water may cause a number of diseases and organ failures when ingested. Excessive copper, Cu(II), intake may alter structure and function of the liver and brain caused by a defect in copper excretion, now known as Wilson disease [1]. Locally, one of the most viable options for treatment of metal ions in wastewater is through adsorption, particularly using montmorillonite (MMT) clay. MMT has a structure that results to an overall slightly negatively charge, which makes it a potentially good adsorbent material [2]. However, modification is still necessary to improve the performance of MMT in removing said trace metals [3].

The adsorption process is defined as the transfer of a contaminant, known as the adsorbate, from the liquid phase to the solid material, known as the adsorbent, either by chemical and/or physical interactions [4]. The process is governed by three main steps: (1) transfer of the metal ion from the bulk liquid on to the surface of the adsorbent, also called the liquid film diffusion; (2) diffusion of the ions within the particle active site, also called intra-particle diffusion; and (3) retention of the metal ions through complexation, precipitation, or physical or chemical sorption [5]. Thus, it is also important to perform modelling to provide insights on the mechanisms involved in a specific adsorption system.

IOP Conf. Series: Materials Science and Engineering **540** (2019) 012004 doi:10.1088/1757-899X/540/1/012004

This study employs Fe(III)-modified montmorillonite (Fe-MMT) to remove Cu(II) ions in synthetic wastewater. Batch adsorption at varying time, initial ion concentration, and temperature will be utilized to fit kinetics, isotherm, and thermodynamic models that will aid in understanding the adsorption mechanism.

2. Methodology

2.1. Synthesis and characterization of Fe-MMT

The Fe-MMT was synthesized based on the methodology of Dela Cruz [6]. Ion exchange of Na-MMT (Southern Clay Products, Inc.) was performed using $FeCl_3$ and NaOH, which was stirred for 12 hours. The obtained particles were dried, grinded, and sieved to particle sizes between 45 μm and 75 μm. The obtained Fe-MMT clay was characterized using a scanning electron microscope with energy dispersive X-ray spectroscopy (SEM-EDX) (Hitachi S-3400).

2.2. Batch adsorption of Cu(II) ions

Synthetic wastewaters at 20ppm concentration were prepared from standard $Cu(NO_3)_2$, with pH maintained at 2.0 to 3.5. An orbital shaker was employed for all the batch experiments, using 50 mL of the aqueous solutions and 30 mg of the Fe-MMT. The clay was allowed to be in contact with the Cu(II) wastewater at varying time intervals until 24 hrs. For experiments involving varying initial concentrations, the Cu(II) were set at 30 ppm, 15 ppm, and 7.5 ppm. Finally, thermodynamic experiments were performed using a multi-stirrer at temperatures between 25°C and 60°C. All ion concentrations before and after adsorption were determined using an inductively coupled plasma-optical emission spectroscopy (ICP-OES, Prodigy7 Teledyne Leeman Labs).

3. Results and discussion

3.1. Morphology of Fe-MMT

SEM micrograph at 3,000x (figure 1) reveal the flaky nature and the wide range of particle size of the synthesized Fe-MMT. The smooth and relatively flat surfaces correspond to the basal plane of the clay. EDX characterization show the elements present in the clay such as silicon (16.96 wt%), iron (9.56 wt%), aluminum (5.03 wt%), magnesium (1.94 wt%), sodium, (0.56 wt%), calcium (0.29 wt%), and oxygen (55.41 wt%). The trace amount of Na suggests that there is an incomplete conversion of some Na-MMT particles to Fe-MMT.

Figure 1. SEM micrograph of Fe-MMT at 3,000x.

IOP Conf. Series: Materials Science and Engineering **540** (2019) 012004 doi:10.1088/1757-899X/540/1/012004

3.2. Adsorption kinetics

The two important parameters observed during adsorption are the %ion removal and the adsorption capacity at any given time (q_t), which may be calculated using equations (1) and (2), respectively. C_0 and C_f are the initial and final concentrations while V is the solution volume and m is the adsorbent mass.

$$\% \text{ ion removal} = \frac{c_o - c_f}{c_o} \times 100\% \tag{1}$$

$$q_t = \frac{(c_o - c_f)V}{m} \tag{2}$$

Adsorption performance of Fe-MMT in removing Cu(II) ions from aqueous solutions is shown in figure 2. There was a steady linear increase in Cu(II) adsorption corresponding to 65.33% removal after 6 hrs. It was then followed by a slower adsorption up to the 12 hr. contact time with 85.86% Cu(II) removed, where it then plateaued and assumed to reach equilibrium. The highest calculated adsorption capacity of Fe-MMT was 34.446 mg/g.

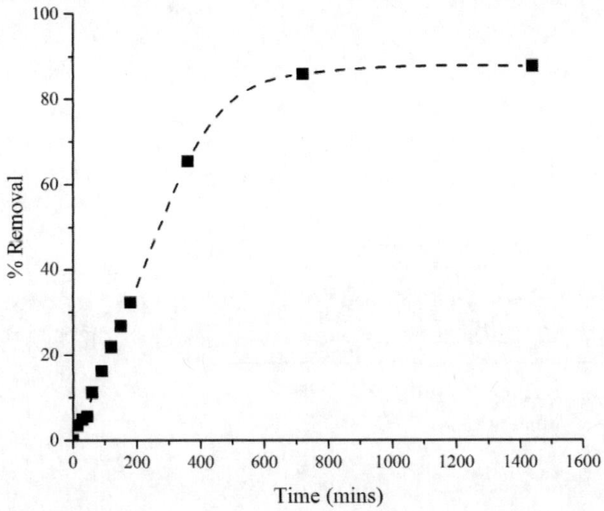

Figure 2. Percent removal of Cu(II) using Fe-MMT at varying time.

Five kinetic models were utilized, namely Lagergren pseudo-first order, pseudo-second order, Elovich, liquid film, and intraparticle kinetic models. The first three models describe the reaction step of the adsorption while the last two differentiates the rate-determining step based on diffusion. Equations of the models are shown in table 1.

Plots of the reaction models are shown in figure 3 while diffusion models are shown in figure 4. Among three reaction models, the highest R^2 value was obtained from the Lagergren pseudo-first order model for Cu(II) adsorption. From the model, the rate constant $k_1 = 0.003 \text{ min}^{-1}$ and $q_e = 35.552$ mg/g were calculated. It is worth noting that the calculated value is close to the experimental value of 34.446 mg/g. Based on the assumptions of the model, the sorption of Cu(II) on Fe-MMT only occurs at specific active sites, and that the Cu(II) ions are adsorbed as a single layer on the surface of the clay [7].

Table 1. Kinetic models and their corresponding equations[1].

Pseudo-first order	$\ln(q_e - q_t) = \ln q_e - k_1 t$
Pseudo-second order	$\frac{t}{q_t} = \frac{1}{k_2 q_e^2} + \frac{1}{q_e} t$
Elovich	$q_t = \beta \ln(\alpha\beta) + \beta \ln t$
Liquid film diffusion	$\ln(1-F) = -k_{fd} t$
Intraparticle diffusion	$q_t = k_i \cdot \sqrt{t}$

[1] Where q_t is the adsorption capacity at any given time, *t*; *qe* is adsorption capacity at equilibrium; *F* is fractional attainment of equilibrium which can be calculated by *qt/qe*; and all the other variables are constants relevant to each kinetic model.

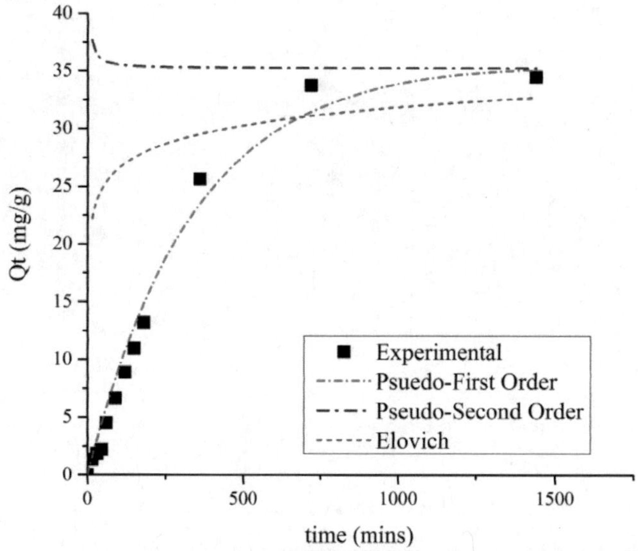

Figure 3. Experimental data with superimposed reaction kinetic models.

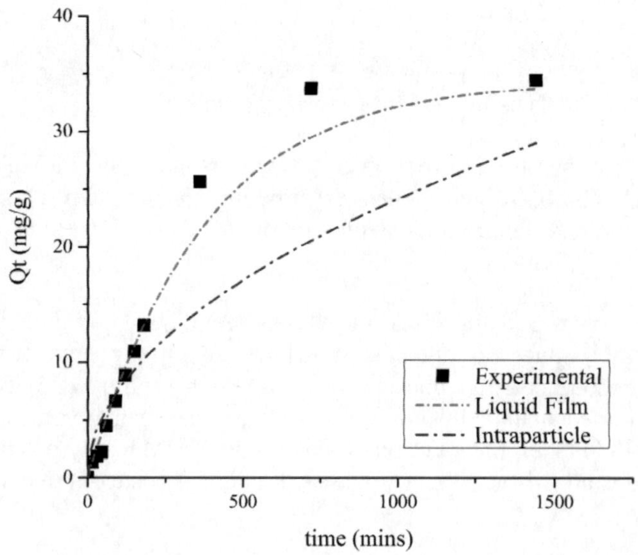

Figure 4. Experimental data with superimposed diffusion kinetic models.

IOP Conf. Series: Materials Science and Engineering **540** (2019) 012004 doi:10.1088/1757-899X/540/1/012004

The diffusion kinetic plots shown in figure 4 were obtained setting the intercept at zero for which it is assumed that there were no adsorbent-adsorbate interactions yet. It can be observed that the Cu(II) adsorption is also governed by the liquid film diffusion model compared to the intraparticle diffusion model. It can thus be inferred that the adsorption process is also rate-determined by the diffusion of the Cu(II) from the bulk liquid to the surface of the Fe-MMT clay.

3.3. Adsorption isotherm modelling

Figure 5 shows the ion removed with respect to increasing ion concentration. A 98.02% removal was observed for Cu(II) at 8 ppm initial concentration, which plateaued at 16 ppm, before finally decreasing to 89.04% at 30 ppm Cu(II). Furthermore, calculations show that increasing the ion concentration also increases the adsorption capacity of the Fe-MMT. The highest adsorption capacity value was obtained at the highest Cu(II) concentration setting, corresponding to 44.945 mg/g.

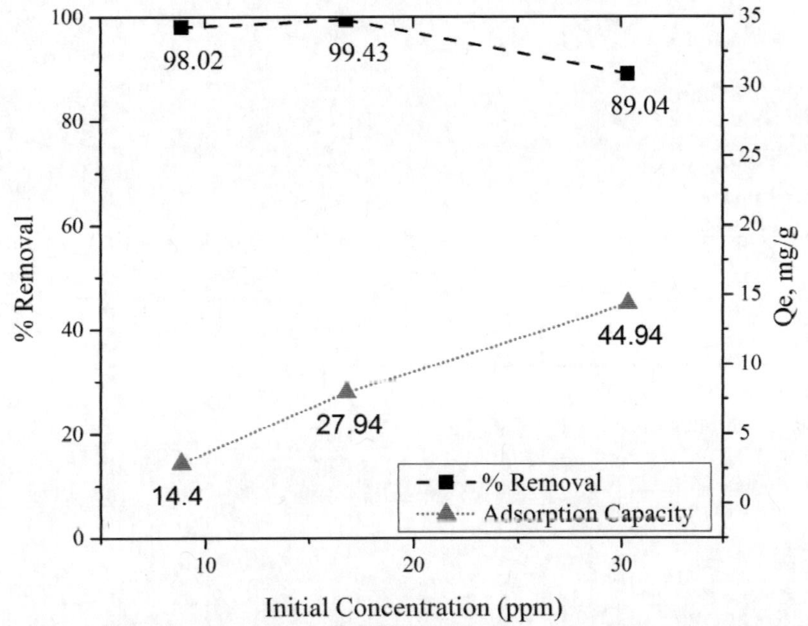

Figure 5. Percent removal of Cu(II) and adsorption capacity of Fe-MMT at varying concentration.

Table 2. Adsorption isotherms models[2]

Langmuir	$\frac{c_e}{q_e}=\frac{1}{K_L}+\frac{a_L}{K_L}c_e;\quad R_L=\frac{1}{1+K_L c_o}$
Freundlich	$\log q_e=\frac{1}{n}\log c_e+\log k_F$
Dubinin-Radushkevich	$\log q_e=\log q_s-\beta\varepsilon^2;\quad \varepsilon=RT\ln\left(1+\frac{1}{c_e}\right);\quad E=\frac{1}{\sqrt{2\beta}}$

[2] Where c_e is the equilibrium ion concentration in the liquid phase; K_L/a_L corresponds to the monolayer adsorption capacity; k_F is constant related to temperature.

Three isotherm models were used in this paper, namely Langmuir, Freundlich, and Dubinin-Radushkevich (D-R) isotherms (see Table 2). It can be observed that from figure 6 that Cu(II) adsorption is better described by the Langmuir isotherm, which suggests that there is a monolayer preferential adsorption onto Fe-MMT. The theoretical monolayer adsorption capacity was calculated to be 48.077 mg/g. Further from the Langmuir model, a dimensionless parameter R_L for each concentration may be calculated using. It is said that adsorption is not favored if $R_L > 1$, linear if $R_L = 1$, favored if R_L is between 1 and zero, and irreversible if $R_L = 0$ [8]. The obtained R_L values are in the range of 0.28 to 0.57 for the implemented settings, thus it may be said that the Cu(II) adsorption is favorable.

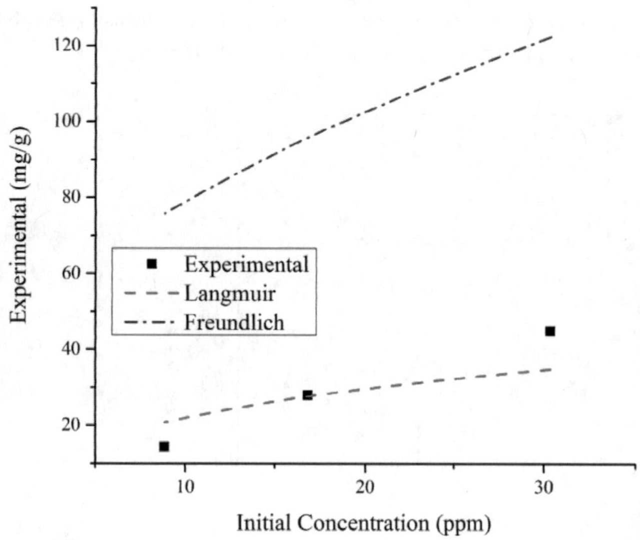

Figure 6. Experimental data with superimposed Langmuir and Freundlich isotherms

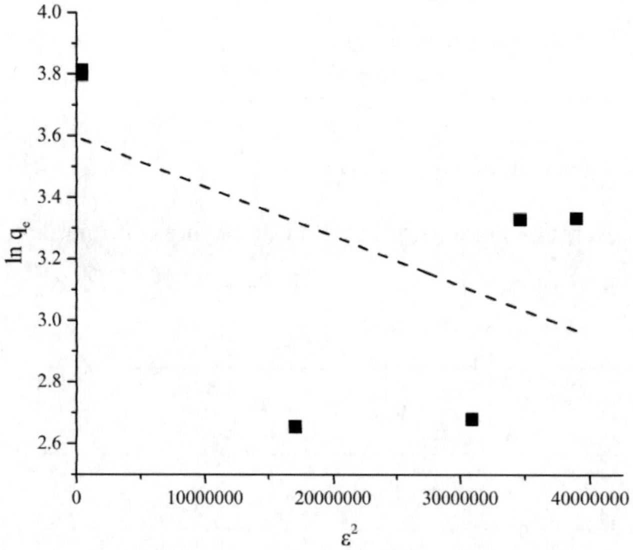

Figure 7. D-R isotherm of Cu(II) adsorption.

From the D-R isotherm (figure 7), the energy (*E*) value for Cu(II) adsorption was calculated to be 5,000 J/mol. Energy values below 8,000 J/mol is attributed to physisorption while E greater than 8,000 J/mol describes chemisorption [9]. Noting that the calculated value falls below the lower limit, it may be concluded that Cu(II) ions adheres to the Fe-MMT surface via physical interactions.

IOP Conf. Series: Materials Science and Engineering **540** (2019) 012004 doi:10.1088/1757-899X/540/1/012004

3.4. Thermodynamic modelling

The adsorption of Cu(II) on Fe-MMT at varying temperature was also explored and shown in figure 8. It can be observed that Cu(II) is affected by the increase in temperature as removal and capacity also increases. This data may be used to calculate the thermodynamic parameters of the Cu(II) adsorption using equation (3). A plot of ln k_d versus 1/T is presented in figure 9 while a summary of the thermodynamic constants is shown in table 2. It can be deduced that the Cu(II) adsorption is an endothermic process with a calculated ΔH° of 23,033.66 J/mol. Furthermore, the ΔS° of the system was found to be 58.18 J/mol K with a corresponding ΔG° at room temperature of 5,696.89 J/mol.

$$\ln K_{eq} = -\frac{\Delta H^o}{RT} + \frac{\Delta S^o}{R} \quad (3)$$

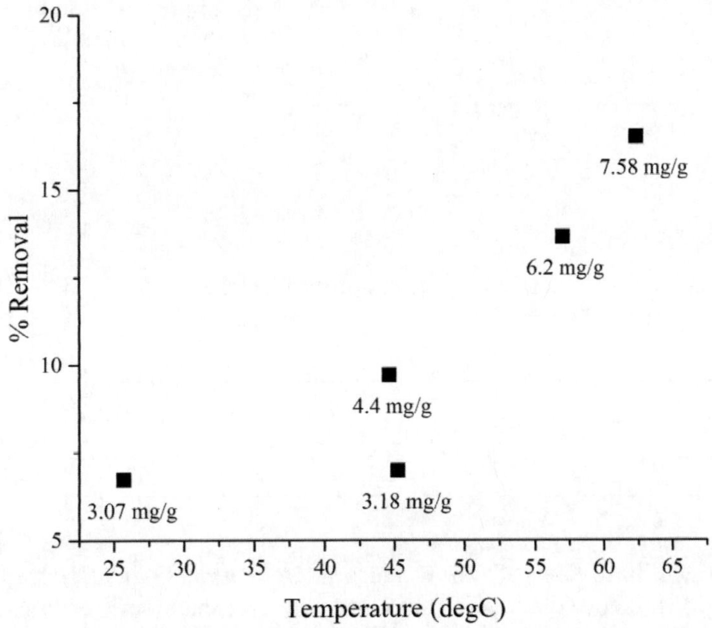

Figure 8. Effect of temperature on Cu(II) adsorption on Fe-MMT.

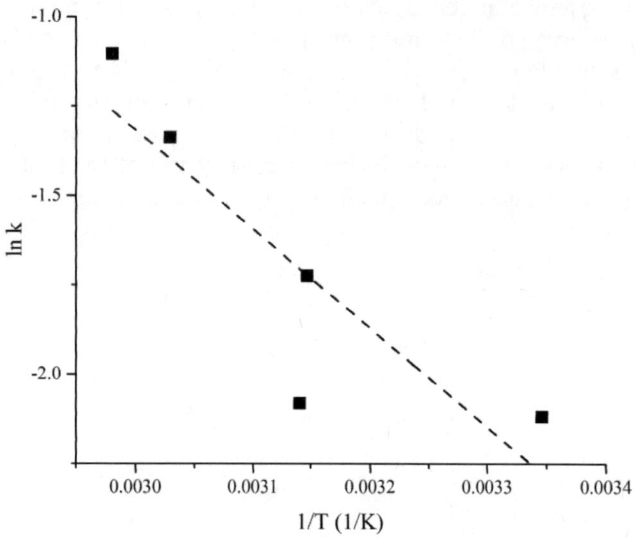

Figure 9. Equilibrium plot to obtain thermodynamic parameters of Cu(II) adsorption

Table 3. Thermodynamic parameters of the Cu(II) adsorption on Fe-MMT.

Temp (K)	*ΔH°* (J/mol)	*ΔS°* (J/mol K)	*ΔG°* (J/mol)
298	23,033.66	58.18	5,696.89
320			4,416.99
335			3,544.33

4. Conclusions

The study successfully utilized Fe-MMT for the adsorption treatment of Cu(II) in synthetic wastewater. The synthesized Fe-MMT showed a layered structure with elemental composition showing successful modification. During kinetic experiments but equilibrium was achieved after 12 hrs with the process being described by the Lagergren pseudo-first order and the liquid film diffusion models. On the other hand, a decrease in percent removal was observed with increasing initial ion concentration. Isotherm modelling showed that monolayer adsorption is apparent for Cu(II) while D-R isotherm showed that the adsorption of Cu(II) on Fe-MMT is via physisorption. Finally, thermodynamic parameters were successfully calculated.

The models and characteristic values obtained in this paper provide valuable insights when Fe-MMT is applied to actual wastewater. It is thus recommended to incorporate the modified clay in other adsorption set-ups that may be more practical for field applications.

References

[1] Uauy R, Maass A and Araya M 2008. Estimating risk from copper excess in human populations *Am J Clin Nutr* **88** 867–71

[2] Krupskaya V, Zakusin S, Tyupina E, Dorzhieva O, Zhukhlistov A, Belousov P and Timofeeva M 2017. Experimental study of montmorillonite structure and transformation of its properties under treatment with inorganic acid solutions *Minerals* **7** 49

[3] Uddin M K 2017. A review on the adsorption of heavy metals by clay minerals, with special focus on the past decade *Chem Eng J* **308** 438–62

[4] Kurniawan T A, Chan G Y S, Lo W H and Babel S 2006. "Physico-chemical treatment techniques for wastewater laden with heavy metals." *Chem. Eng J* **118** 83–98.

[5] Ijagbemi C O, Baek M H and Kim D S 2010. Adsorptive performance of un-calcined sodium exchanged and acid modified montmorillonite for Ni2+ removal: Equilibrium, kinetics, thermodynamics and regeneration studies." *J Hazard Mater* **174** 746–55.

[6] Dela Cruz M L, Arano K, Dela Pena E M and Diaz L J 2013. "Nanoclay-supported zero-valent iron as an efficient adsorbent material for arsenic." *Adv Mater Res* **686** 296–304

[7] Largitte L and Pasquier R 2016. "A review of the kinetics adsorption models and their application to the adsorption of lead by an activated carbon." *Chem Eng Res Des* **109** 495–504

[8] Bessbousse H, Rhlalou T, Verchère J-F and Lebrun L 2008. "Sorption and filtration of Hg(II) ions from aqueous solutions with a membrane containing poly(ethyleneimine) as a complexing polymer." *J Membr Sci* 997–1006

[9] Demiral H and Güngör C 2016. "Adsorption of copper (II) from aqueous solutions on activated carbon prepared from grape bagasse." *J Clean Prod* **124** 103–13.

Acknowledgements
The authors acknowledge the support of DOST PCIEERD through the Mineral Extraction with Responsibility for Sustainability (MinERS) Project E, and the UPD OVCRD through the Thesis and Dissertation Grants. JK Cruz further acknowledges the Crisostomo A. Ortigas Professorial Chair. Special thanks are given to Ms. Nerissa Quenga and Kayvee Surop for aiding in the ICP runs.

IOP Conf. Series: Materials Science and Engineering **540** (2019) 012005 doi:10.1088/1757-899X/540/1/012005

Kinetics and Isotherms Studies of the Adsorption of Hg(II) onto Iron Modified Montmorillonite/Polycaprolactone Nanofiber Membrane

Lester Raj Somera, Ralph Cuazon, John Kenneth Cruz and Leslie Joy Diaz

Department of Mining, Metallurgical and Materials Engineering, University of the Philippines Diliman, Quezon City, Philippines
E-mail: lasomera@up.edu.ph

Abstract. Hg(II) ions from mining and urban runoffs can easily contaminate various water sources due to their high solubility in water. Due to its simplicity and viability, adsorption was utilized to treat Hg(II)-contaminated water. An adsorbent was fabricated by immobilizing Fe-modified montmorillonite particles in polycaprolactone through electrospinning. The SEM images of the fabricated membrane showed a nanofiber network with fiber diameter ranging from 96 nm to 569 nm and an average of 198 nm. The adsorption behavior of the fabricated membrane was investigated using kinetics and isotherm models. It was found that the adsorption of Hg(II) ions onto the membrane follows a pseudo-second order model and was best described by the Freundlich isotherm. The adsorption capacity was found to increase with increasing contact time and with increasing initial Hg(II) concentration, with the highest adsorption capacity of 14.25 mg/g. The sorption energy was calculated using the Dubinin-Kaganer-Radushkevich isotherm, which obtained a value of -2.68 $kJmol^{-1}$ suggesting that the type of adsorption was physisorption.

1. Introduction

Dissolved mercury, Hg(II), is considered to be one of the most toxic heavy metals. The toxicity of mercury has been extensively studied and its concentrations in water sources are constantly monitored [1]. There are several methods to treat mercury laden waste water, however most are not preferred due to high costs, high energy consumption and difficulty of use [2] [3]. Among the available methods, adsorption is proven to be one of the most economical and efficient for removing heavy metals [4].

Adsorption is the adhesion of ions or molecules onto a solid surface. Several mathematical models have been proposed to describe the kinetics of adsorption. These models are used to describe the kinetic process of adsorption and the adsorption mechanism. These can generally be classified as adsorption reaction models (pseudo-first order rate equation, pseudo-second order rate equation, and Elovich equation) and adsorption diffusion models (liquid film diffusion model, and intra-particle diffusion model) [5]. Adsorption kinetics follows a three step mechanism: (i) film diffusion or the transfer of solute to the adsorbent surface (ii) intra-particle diffusion or the transfer of solute from the adsorbent surface to the intra-particle adsorption sites, and (iii) mass action or the retention of solute on the active site via adsorption complexation or precipitation [6]. The rate-limiting step, which can determine the overall adsorption rate of the system for the adsorption process, can be any of the three steps or a combination of two [7]. The linearized equations of the relevant adsorption reaction models and adsorption diffusion models are listed in table 1.

Table 1. Linearized equations of adsorption reaction and adsorption diffusion models.

Kinetic model	Linearized equation	
Lagergren pseudo-first order equation	$\log(q_e - q_t) = log q_e - \frac{k_1 t}{2.303}$	(1)
Pseudo-second order equation	$\frac{t}{q_t} = \frac{1}{k_2 q_e^2} + \frac{t}{q_e}$	(2)
Elovich equation	$q = \alpha \ln(a\alpha) + \alpha \ln(t)$	(3)
Liquid film diffusion	$\ln\left(1 - \frac{q_t}{q_e}\right) = -R^1 t$ $R^1 = \frac{3D_e^1}{r_o \Delta r_o k}$	(4)
Intra-particle diffusion	$q_t = k_{int} \cdot \sqrt{t}$	(5)

Similarly, there are several isotherms that can be used to describe adsorption. These isotherms are used to correlate the equilibrium adsorption data with the mathematical models describing kinetics. It describes the adsorbent-adsorbate interaction and is critical in optimizing the use of adsorbents [8]. The Langmuir and Freundlich isotherms are the two most commonly used isotherm model for adsorption. While the Dubinin-Kaganer-Radushkevich isotherm is commonly used to compute the sorption energy of the sorption process. The linearized forms of the relevant adsorption isotherms are shown in table 2.

Table 2. Linearized equations of isotherm models.

Isotherm model	Linearized equation	
Langmuir isotherm	$\frac{1}{q_e} = \frac{1}{K_a q_m} \frac{1}{C_o} + \frac{1}{q_m}$	(6)
Freundlich isotherm	$log q_e = K_F + \frac{1}{n} log C_e$	(7)
Dubinin-Kaganer-Radushkevich isotherm	$lnQ = lnQ_{max} - \beta\varepsilon^2$ $\varepsilon = RTln\frac{1}{C_e}$	(8)

The objective of this study is to determine the kinetic model and adsorption isotherm that best describes the adsorption of Hg(II) onto the Fe-MMt/PCL nanofiber membrane. To achieve this, adsorption capacity was monitored as a function of contact time, and metal ion concentration.

2. Methodology

2.1. Fe-MMt clay preparation

The method used for the preparation of Fe-MMt clay was adapted from the study of [9] Nanofil 116 was dispersed distilled water while $FeCl_3$ solution was being added dropwise during dispersion. After which, additional amounts of distilled water and NaOH solution was added to the solution. The resulting solution was stirred at 500 rpm for 12 hours. It was then filtered, collecting only the residue which is the Fe-MMT clay. The residue collected was oven dried at 80 °C for 5 hours. The dried residue was ground and sieved, collecting only particles less than 45 μm and then ball milled for 2 hours at 100 rpm.

2.2. Fe-MMt/PCL nanofiber membrane fabrication

Fe-MMt clay and Cloisite 24A were dispersed in dichloromethane (DCM) solution. This mixture was sonicated for 30 minutes, and added with polycaprolactone (PCL) pellets. The new mixture was sonicated for 1 hour, and was shaken every 5 minutes for the last 30 minutes. This solution was electrospun at a voltage of 33 kV, spinning distance of 13 cm, and barrel collector speed of 450 rpm using an *Inovenso Nanospinner 24*. The process was done three times to fabricate a single sheet composed of Fe-MMt/PCL nanofiber membrane. The fabricated membrane was cut into circular plates with a 58 mm diameter.

2.3. Adsorption experiments

Three different initial concentrations (10, 20, and 30 ppm) of Hg(II) ions at pH 3 were prepared. Adsorption experiments were conducted using a modified setup adapted from [10]. One liter of the Hg(II) solution was allowed to pass through the Fe-MMt membrane repeatedly for a total of 360 minutes. The changes in concentration were monitored by collecting 15 mL aliquots at 0, 15, 30, 45, 60, 120, 240, and 360 minute intervals during the adsorption duration.

2.4. Characterization

The fabricated membrane was viewed under a *Hitachi SU8230 Scanning Electron Microscope (SEM)* to examine its surface morphology. The fiber diameter was measured using ImageJ© software. EDS analysis was subsequently performed on the membrane surface to confirm the presence of Fe-MMt particles. Meanwhile, the concentration of the aliquot collected for each time interval of the adsorption experiment was determined using *X-Ray Fluorescence Spectroscopy (XOS HDXRF).* The collected data was analyzed using OriginPro© to plot the linearized equations of the relevant kinetic models and isotherms.

3. Results and discussion

3.1 Morphology of Fe-MMt/PCL nanofiber membrane

Fe-MMt particles were immobilized in PCL through electrospinning. Figure 1 shows the resulting membrane made up of composite nanofibers.

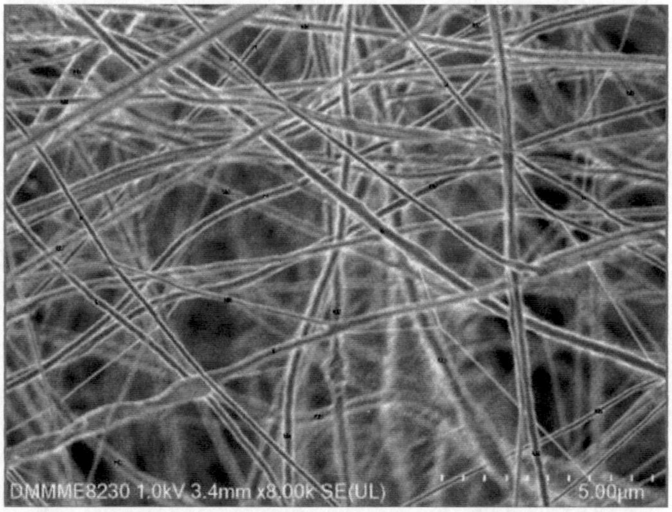

Figure 1. Morphology of Fe-MMt Membrane at 8000x.

The measured fiber diameter in figure 1 was found to be within the 96-569 nm range with an average fiber diameter of 198 nm. EDS analysis of exposed clay particles confirmed the presence Fe, Si, Mg, Al, and O. This suggests that the exposed clay paticles are indeed Fe-MMt.

3.2 Adsorption kinetics

The adsorption of Hg(II) ions onto Fe-MMt/PCL nanofiber membrane has been investigated as a function of time. Changes in the ion concentration were monitored at specific intervals from 0 to 360 minutes. Figure 2 shows the relationship between the adsorption capacity and contact time at three different initial concentrations of Hg(II): 10 ppm, 20 ppm, and 30 ppm.

IOP Conf. Series: Materials Science and Engineering **540** (2019) 012005 doi:10.1088/1757-899X/540/1/012005

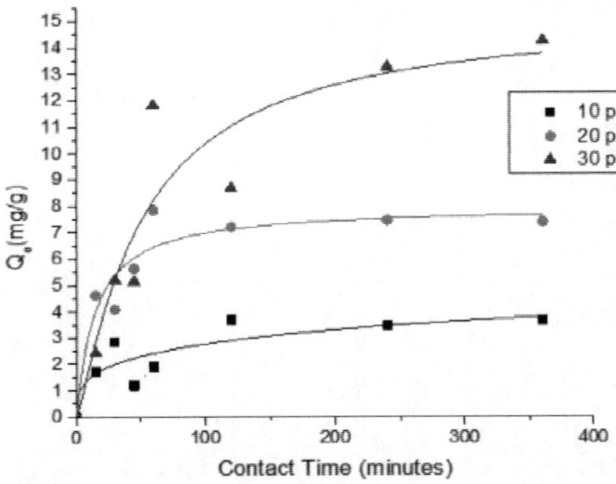

Figure 2. Adsorption Capacity Curve of Hg(II) ions.

Generally, it was observed that for all settings, the adsorption capacity increases with increasing contact time until no significant change in the adsorption capacity can be observed. A high initial removal was observed from the initial contact time up to 100 min. A slower removal followed until a plateau on the plot was observed, indicating equilibrium. This may be attributed to the high availability of adsorption sites at the start of adsorption. However, as adsorption progresses, the sites are gradually filled and ions have to intensely compete to be adsorbed [11]. The data collected from the adsorption as a function of time was used to determine the adsorption kinetic model and the rate limiting step of the Hg(II) adsorption. Figures 3-7 below show the linearized kinetic model plots for the adsorption of Hg(II) ions:

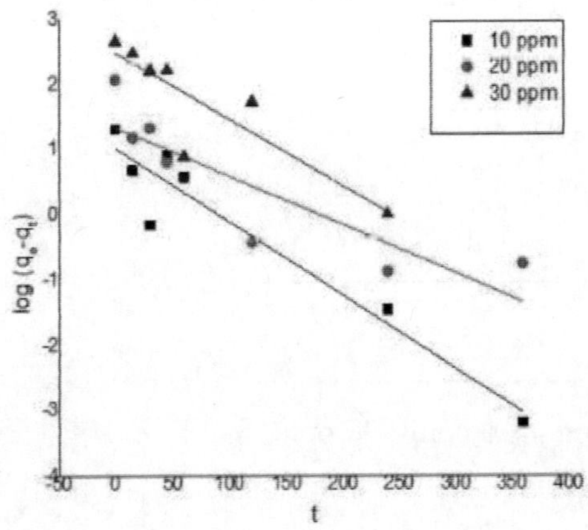

Figure 3. Linearized Lagergren pseudo-first order equation plot for Hg(II) adsorption.

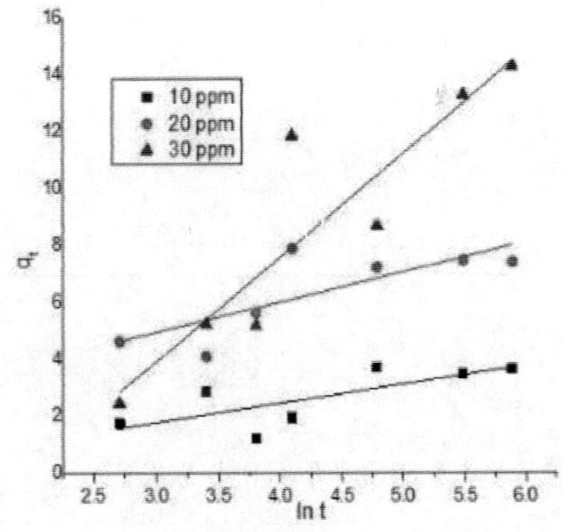

Figure 4. Linearized Elovich equation plot for Hg(II) adsorption.

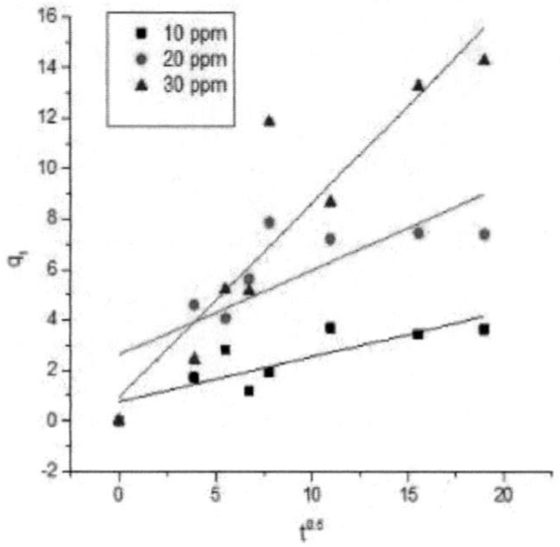

Figure 5. Linearized intra-particle diffusion model plot for Hg(II) adsorption.

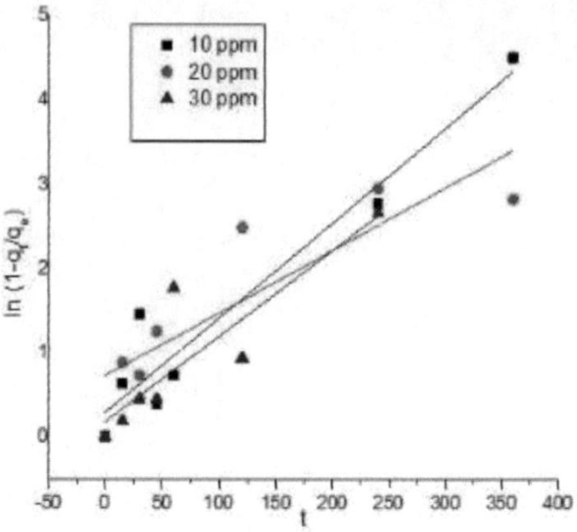

Figure 6. Linearized liquid film diffusion model plot for Hg(II) adsorption.

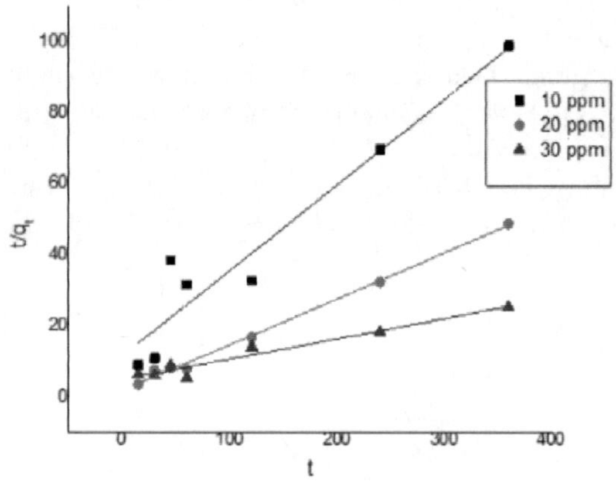

Figure 7. Linearized pseudo-second order equation plot for Hg(II) adsorption.

Table 3. Summary of kinetic models correlation coefficients

Kinetic Model	R^2		
	10 ppm	20 ppm	30 ppm
Lagergren Pseudo-first order	0.929	0.764	0.772
Pseudo-second order	0.929	0.995	0.949
Elovich	0.558	0.621	0.827
Intra-particle diffusion	0.702	0.621	0.840
Liquid film diffusion	0.929	0.764	0.772

R^2 values for the pseudo-second order were generally higher than other kinetic models at different concentrations as shown in table 3; thus, adsorption of Hg(II) ions onto the Fe-MMt/PCL nanofiber membrane can be said to follow a pseudo-second order kinetics. Rate constant, k_2, was variable with no apparent trend at the different initial Hg(II) concentrations. In addition, table 4 below compares the calculated q_e to experimental q_e.

Table 4. Equilibrium adsorption capacity q_e

Hg(II) concentration, ppm	Calculated q_e, mg/g	Experimental q_e, mg/g	%deviation
10	4.15	3.69	11.1
20	7.73	7.86	1.68
30	17.42	14.25	18.2

The pseudo-second order model is usually used to describe adsorption where the surface-reaction kinetic step controls the overall adsorption rate. The reaction order for this model is two with respect to the number of adsorption sites which means that each adsorbed molecule is attached to two active sites of the adsorbent [7].

3.3 Adsorption isotherms

The adsorption of Hg(II) ions onto Fe-MMt/ PCL nanofiber membrane has been investigated as a function of initial concentration. It was observed in figure 8 that maximum adsorption capacity increases with increasing initial Hg(II) concentration.

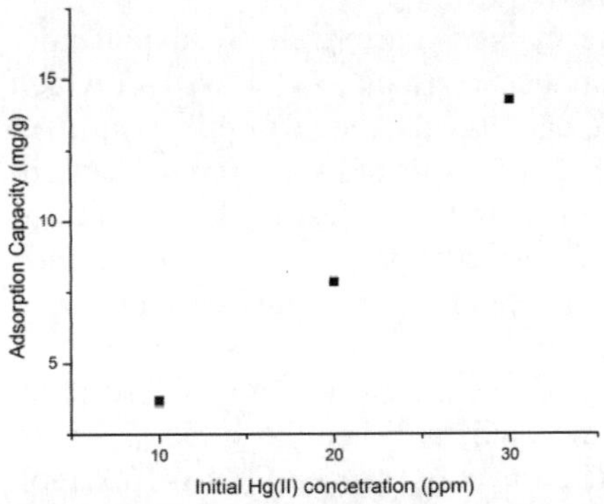

Figure 8. Adsorption capacity as a function of initial concentration.

Increasing the initial metal ion concentration increases the driving force to overcome the mass transfer resistance that hinders the movement of the metal ions from the bulk solution to the clay surface. This allows for more metal ions to come into contact with the adsorbent particles [12]. As a result, the adsorption capacity of the membrane increases as the initial metal ion concentration increases. This can be observed in figure 8, where the highest adsorption capacity of 14.254 mg/g was obtained for the adsorption of the 30 ppm Hg(II) solution.

The data obtained from the adsorption of Hg(II) as a function of initial concentration was analyzed in terms of the Langmuir, the Freundlich, and the Dubinin-Kaganer-Radushkevich isotherms as shown in figures 9, 10, and 11, respectively.

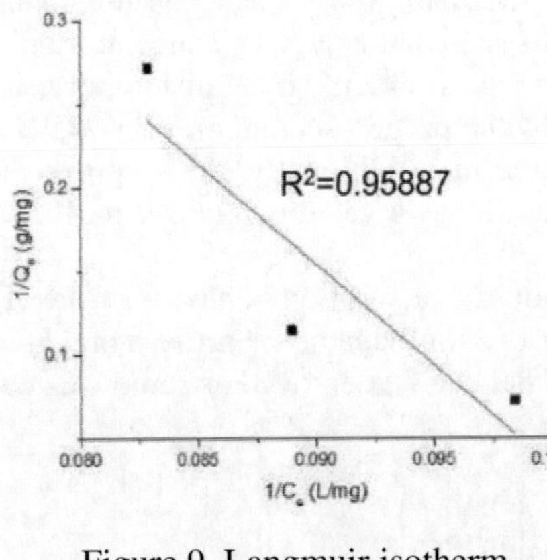

Figure 9. Langmuir isotherm.

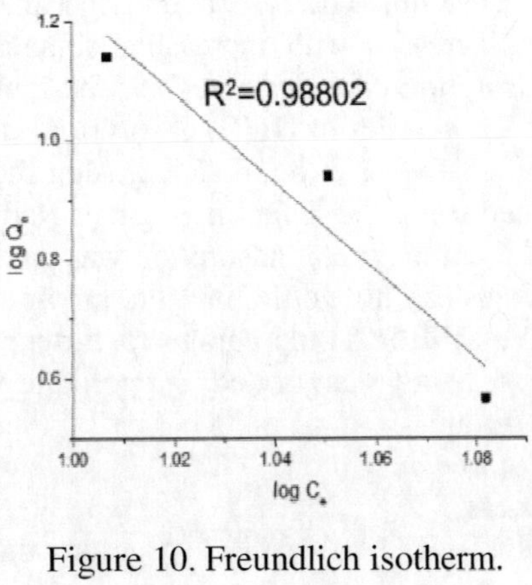

Figure 10. Freundlich isotherm.

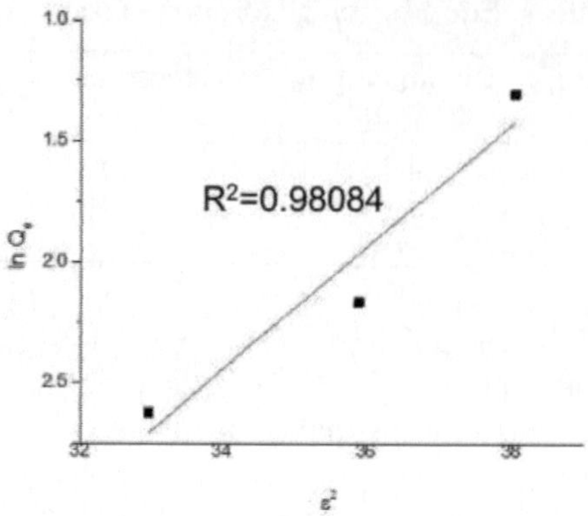

Figure 11. Dubinin-Kaganer-Radushkevich isotherm.

Based on the values of the correlation coefficients (R^2) obtained, it was found that the Freundlich Isotherm best describes the adsorption behavior of Hg(II) on Fe-MMt/PCL. A previous study on Hg(II) adsorption onto montmorillonite also followed a Freundlich isotherm [13].
The multilayer adsorption described by the Freundlich isotherm is characterized by exponential distribution of heterogeneous sites [14]. The value of the Freundlich constant, K_F, is 0.0103 which suggests a low degree of heterogeneity. Another parameter, 1/n, can classify the adsorption isotherm as either concave or convex depending on its value. A value of $(1/n) < 1$ denotes a convex isotherm where sorption energy is inversely proportional with surface concentration. Meanwhile, a value of $(1/n) > 1$ describes a concave isotherm where sorption energy is directly proportional with surface concentration [15]. The obtained value for 1/n is 2.3795 which is greater than unity thus, an increase in surface concentration increases the sorption energy.
The Dubinin-Kaganer-Radushkevich (DKR) isotherm was used to determine the value of the sorption energy E through equation 9, which can confirm the adsorption mechanism. When the value of E is in the range of -1 to -8kJmol^{-1}, it is indicative of physisorption. When it is in the range of -8 to -16 kJmol^{-1} it is indicative of ion-exchange. While, when it is in the range of -20 to -40 kJmol^{-1} it is indicative of chemisorption [14]. The E value obtained for Hg(II) is -2.6803 kJmol^{-1}, well within -1 to -8 kJmol^{-1} which shows that the type of adsorption for Hg(II) ions is physisorption.

$$E = -\frac{1}{\sqrt{-2\beta}} \tag{9}$$

4. Conclusion

Fe-MMt/PCL nanofiber membrane was successfully fabricated through electrospinning which produced a nanofiber network with fiber diameter ranging from 96 nm to 569 nm and an average fiber diameter of 198 nm. The adsorption experiments showed that the adsorption capacity of the nanofiber membrane increases with increasing contact time, and with increasing Hg(II) concentration. The highest adsorption capacity noted for the fabricated membrane was 14.25 mg/g. Furthermore, it was found that the kinetics of Hg(II) adsorption can be described by the pseudo-second order model and its adsorption behavior can be modeled using the Freundlich isotherm. Additionally, the sorption energy E, calculated using the Dubinin-Kaganer-Radushkevich isotherm, has a value of -2.68 kJmol^{-1}, which confirmed that the type of adsorption was physisorption.
Future studies can utilize the insights provided by this study on the adsorption behavior of the Hg(II) onto the Fe-MMt/PCL membrane to further improve its adsorption capacity. Furthermore, the Fe-MMt/PCL may be incorporated in a portable filtration set-up that can be deployed as rapid wastewater treatment set-up.

5. References

[1] WHO W H O 2013 WHO | Mercury and health *Fact Sheet n 361* 1

[2] Wang G, Hua Y, Su X, Komarneni S, Ma S and Wang Y 2016 Cr(VI) adsorption by montmorillonite nanocomposites *Appl. Clay Sci.* **124–125** 111–8
[3] Dawlet A, Talip D, Mi H Y and MaLiKeZhaTi 2013 Removal of Mercury from Aqueous Solution Using Sheep Bone Charcoal *Procedia Environ. Sci.* **18** 800–8
[4] Karnib M, Kabbani A, Holail H and Olama Z 2014 Heavy metals removal using activated carbon, silica and silica activated carbon composite *Energy Procedia* vol 50 (Elsevier) pp 113–20
[5] Qiu H, Lv L, Pan B, Zhang Q, Zhang W and Zhang Q 2009 Critical review in adsorption kinetic models *J. Zhejiang Univ. A* **10** 716–24
[6] Ijagbemi C O, Baek M-H and Kim D-S 2009 Montmorillonite surface properties and sorption characteristics for heavy metal removal from aqueous solutions *J. Hazard. Mater.* **166** 538–46
[7] Plazinski W, Rudzinski W and Plazinska A 2009 Theoretical models of sorption kinetics including a surface reaction mechanism: A review *Adv. Colloid Interface Sci.* **152** 2–13
[8] Ali Fil B 2016 Particulate Science and Technology An International Journal Isotherm, kinetic, and thermodynamic studies on the adsorption behavior of malachite green dye onto montmorillonite clay *Part. Sci. Technol.* **34** 118–26
[9] Dela Cruz M L 2011 *Adsorption Mechanism of Arsenic Ions on Iron-Modified Montmorillonite* (University of the Philippines - Diliman)
[10] Dela Pena E M B 2011 *Dynamic adsorption of aqueous arsenic on electrospun nanofibrous mat* (University of the Philippines-Diliman)
[11] Benhammou A, Yaacoubi A, Nibou L and Tanouti B 2005 Adsorption of metal ions onto Moroccan stevensite: Kinetic and isotherm studies *J. Colloid Interface Sci.* **282** 320–6
[12] Soleimani M and Siahpoosh Z H 2015 Ghezeljeh nanoclay as a new natural adsorbent for the removal of copper and mercury ions: Equilibrium, kinetics and thermodynamics studies *Chinese J. Chem. Eng.* **23** 1819–33
[13] Green-Ruiz C 2005 Adsorption of mercury(II) from aqueous solutions by the clay mineral montmorillonite *Bull. Environ. Contam. Toxicol.* **75** 1137–42
[14] Eba F, Gueu S, A-Mvongbote E, Ondo J A, Yao B K, Nlo J N and Biboutou R K 2010 Evaluation of the absorption capacity of the natural clay from Bikougou (Gabon) to remove Mn (II) from aqueous solution . *Int. J. Eng. Sci. Technol.* **2** 5001–16
[15] Can N, Ömür B C and Altındal A 2016 Modeling of heavy metal ion adsorption isotherms onto metallophthalocyanıne film *Sensors and Actuators B: Chemical***237**953–61

Acknowledgements
We would like to acknowledge the MinERS Project E for the initiative to develop innovative materials for treating small-scale mining waste, which made the way for the creation of this study. We also would like to acknowledge the financial support provided by the DOST-PCIEERD.

IOP Conf. Series: Materials Science and Engineering **540** (2019) 012006 doi:10.1088/1757-899X/540/1/012006

Magnetic Carbon Nanofibers from Horse Manure via Hydrothermal Carbonization for Methylene Blue Adsorption

Napat Kaewtrakulchai[1], Ampol Putta[1], Warit Pasee[1], Kajornsak Fuangnawakij[2], Gasidit Panomsuwan[3], Apiluck Eiad-ua [1]

1 College of Nanotechnology, King Mongkut's Institute of Technology Ladkrabang, Bangkok, Thailand

2 National Nanotechnology Center (NANOTEC), National Science and Technology Development Agency, Thailand Science Park, Patumtani, Thailand

3 Department of Materials Engineering, Faculty of Engineering, Kasetsart University, Bangkok, Thailand

E-mail: apiluck.ei@kmitl.ac.th

Abstract. Disposal and recycle of waste biomass is of great concern. Thermochemically converting the waste biomass to carbon nanomaterials is an interesting because of environmental friendly, low cost and local availability. In this work, magnetic carbon nanofibers have been synthesized by hydrothermal and carbonization of the magnetite preloaded on horse manure which is controllable temperature and additive of catalyst. It was found that Fe is able to form magnetic carbon nanofibers (M-CNFs). Furthermore, magnetic carbon nanofibers were used as an adsorbent for methylene blue adsorption. Synthesized magnetic sorbents exhibited high performance on methylene blue adsorption and it is successfully separated from the water by magnetic separation.

1. Introduction

The manures from livestock sections have been traditionally used for increase the crop production as an organic fertilizer providing high organic nutrients, nitrogen, phosphorus and potassium. However, animal manure utilization reveals the negative impacts to surrounding environment. The air pollution as an odour nuisance is the main influence on environmental concerns. On one hand, the organic substances such as carbon, nitrogen, hydrogen, etc. compounded of livestock manure can be converted to a high value products. Based on energy production, the production of biogas form manure has been widely investigated recently because of energy and environment issues concerns [1]. Biogas production is employed using anaerobic digestion process can be used in several uses for heat production such as cooking in household, direct combustion to generate electricity [2, 3]. On the other hand, carbon material produced from organic materials has received a much attention since it is abundance of carbon source. Hence, it become a favourable raw feedstock for carbon material production due to alternative precursor with huge availability in nature [4-6]. The development of animal manure to a carbon materials such as biochar, activated carbon, carbon fiber, etc. is extensively effective due to its potential such as the reduction of greenhouse gases emission, cost effective solution, and carbon sequestration [7]. Animal manures developed to porous carbons have some significances such as adsorption performance, high mechanical property, high thermal stability for novel applications. It was widely applied in the fields of environmental technologies such as adsorbent and soil amender. Also, energy utilizations such as electronic devices, catalyst supporter and even

alternate solid fuel [8]. For these reasons, the several technologies have been applied for its manufacturing.
Currently, several thermochemical conversion technologies have been applied for the carbon material manufacturing [9]. Hydrothermal treatment and carbonization is the most effective thermal techniques for carbon production from organic substances [10]. In hydrothermal treatment, the chemical structure of biomass materials have been developed by temperature and pressure during the reaction leads to high carbon content and rising in hydrophobic property obtaining at relatively low temperature (140-260 °C) [11]. While, the carbonization provides the pore formation and also increases in carbon percentage from devolatilization during high temperature operation [12]. Carbonization of organic substances to carbon is normally conducted at moderate temperature ranges between 400 to 900 °C. Vast recent studies reported the process variables such as treatment temperature, holding time in HT process has a significant impact on the specific surface area. A small number of micropores results in a small surface area is only generated from HT process compared to activated carbon. Also, the pores of HT by-product can be blocked via the repolymerization of water soluble compound at long residence time conditions leads to the decrease of specific surface area. In this respect, further carbonization at higher temperatures, the surface areas could be elevated dramatically due to an increase in micropore structure. The HT process combined carbonization can be suitable for the production of adsorbent or catalyst supporter [13-15]. In this study, horse manure (HM) is a famous residue with natural fibrous characteristics was selected as a raw feedstock for synthesis of magnetic carbon nanofibers (M-CNFs) using hydrothermal treatment combined carbonization. The effects of magnetic precursors (*i.e.* iron nitrite and iron oxide) has been investigated on 10 wt.%. of magnetic precursor using various impregnation ratios of 10:0 to 0:10 (iron nitrate and iron oxide, wt./wt.) to optimize and serve the magnetic characteristic. The prepared M-CNFs were comprehensively characterized by scanning electron microscope (SEM) and fourier transform infrared spectroscopy (FTIR), X-ray diffractometer (XRD) and vibrating sample magnetometer (VSM), respectively. The produced M-CNFs produced from optimum condition was tested to remove methylene blue. Also, the M-CNFs is successfully abled to separate by magnetic characteristic.

2. Experimental

2.1 Magnetic carbon nanofibers preparation

Horse manure (HM) were collected as raw precursor for synthesis of magnetic carbon nanofibers (M-CNF). Before the experiments, the feedstocks were crushed and sieved into the size of 800 μm. In M-CNF synthesis experiment, M-CNF was completely obtained via two-step process which is a hydrothermal treatment followed by carbonization. 30 g prepared horse manure was mixed with 60 ml deionized water. Then, the suspension was transferred into a stainless steel autoclave and it was directly heated in an electrical oven at constant temperature of 200 °C under 24 h controlled holding time to reach hydrothermal treatment which has been reported elsewhere [7]. Afterwards, the effects of iron precursor types and impregnation ratios on the magnetic feature of M-CNF were comprehensively investigated. the solid char from HT process was impregnated with 10 wt.% of magnetic precursor using a various impregnation ratios of 10:0 to 0:10 (iron nitrate and iron oxide, wt./wt.) to optimize and serve the magnetic characteristic. The impregnated samples were dried at 90 °C for overnight before carbonization. Finally, carbonization was conducted in a tubular furnace system, the samples were carbonized at the temperature of 900 °C using a ramping rate of 10 °C/min for 2 h and then cooled to ambient under the nitrogen surrounding. The prepared M-CNF was washed several times with deionized water for cleaning up impurities and then dried and kept for further characterization. The physical morphology and surface functional characteristics of as-prepared M-CNFs were characterized by Scanning Electron Microscope (SEM) and Fourier Transform Infrared Spectroscopy (FTIR), respectively. While, X-ray diffractometer (XRD) was analyzed to reveals the crystallography of as-prepared M-CNF. While, the magnetic spectra was determined using vibrating sample magnetometer (VSM).

2.2 Methylene blue adsorption study

IOP Conf. Series: Materials Science and Engineering **540** (2019) 012006 doi:10.1088/1757-899X/540/1/012006

The adsorption studies were conducted in 100 mL flasks. The methylene blue (MB) solution of 30 mL with the different initial concentrations 25 – 100 mg/L used as an adsorbate were placed in this flask. Then, 0.1 g of as-prepared M-CNFs was added into the MB solution and each sample was kept in an isothermal shaker (30 °C) for 12 h to reach equilibrium of the solution. Another flask containing the same concentration of MB solution without M-CNFs was set to analyze as a blank. Then, the final concentration of MB solutions were directly filtered and analyzed rapidly. The amount of adsorption at equilibrium, Q_e (mg/g), was calculated by:

$$Q_e = (C_0\text{-}C)V / W \qquad (1)$$

Table 1. Proximate and elemental analysis of horse manure.

Parameter	
Proximate analysis (wt.%)	
Moisture	7.79
Ash	13.54
Volatile matter	63.34
Fixed carbon*	15.33
Ultimate analysis (wt.%)	
C	45.9
H	6.4
N	1.0
O*	35.8

**calculated by different*

where C_0 and C (mg/L) are the liquid-phase concentrations of dye at initial and equilibrium, respectively. V is the volume of the solution (L) and W is the mass of dry adsorbent used (g).

3. Results and discussion

3.1. M-CNFs characteristic

3.1.1 Chemical composition of M-CNFs

Proximate analysis, was analyzed followed standard technique, [16] demonstrated the chemical composition of biomaterials which composes of moisture, volatile matter, fixed carbon and ash. While, the ultimate analysis was used to determine percentage of carbon, hydrogen using CHN elemental analyzer [17]. The oxygen was calculated by the different of 100 percent. The result on proximate analysis of horse manure shows in table 1. It found that horse manure contains 15.33 wt.% of fixed carbon content and ash content about 13.54 wt.%. Furthermore, moisture is lower than 8.0% in weight. The volatile matter content is approximately 64 wt.%. In addition, elemental analysis carried out to examine the C, H, N, O elemental composition. The ultimate analysis obtained on C, H, N, O is approximately 45.9%, 6.4%, 1.0%, 35.7%, respectively.

3.1.2 SEM micrographs of M-CNFs

Physical morphology of M-CNFs samples was observed by scanning electron microscope at the magnificent of 1000X. Figure 1. (A) displays the surface of raw horse manure. Whereas, Figure 1. (B)-(L) displayed SEM images of MCNF samples produced by hydrothermal processes at 200 °C for 24 hours and carbonization at 900 °C for 2 h, and add the iron nitrate and iron oxide ($Fe(NO_3)_3$: Fe_2O_3) in a ratio of (0:10) to (0:10). The rough surface and cavities characteristic were found because of high temperature process during pyrolysis. The volatile substances was completely decomposed leaves the carbon matrix representing the pore formation [17, 18]. It is also noticeable that the iron oxide distribution on the surface of the carbon causes the carbon to be magnetized.

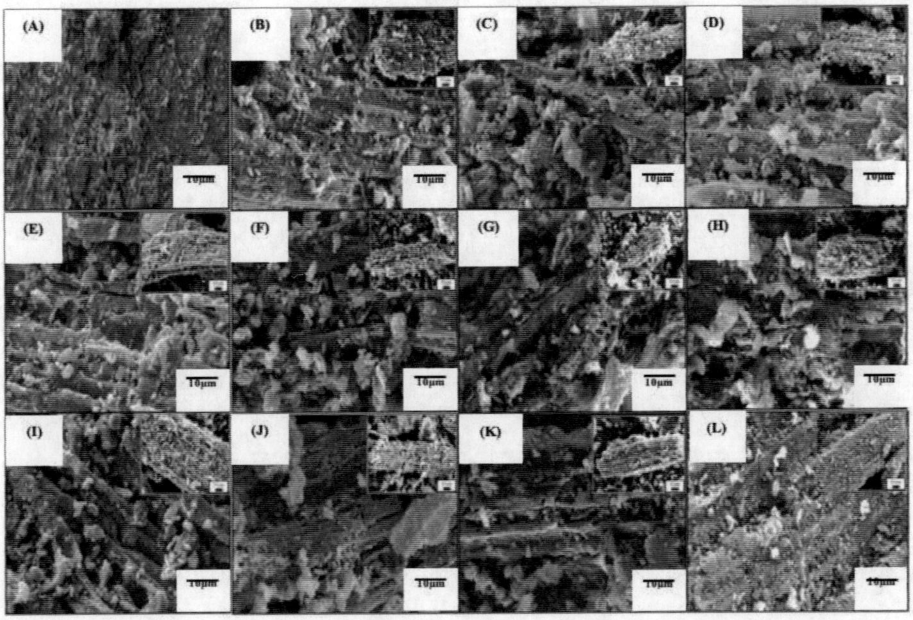

Figure 1. Scanning Electron Microscope (SEM) of (1000x) and (300x) (A). horse manure, the horse manure through hydrothermal processes at 200 °C for 24 h and carbonization at 900 °C for 2 h, and add the iron nitrate and iron oxide(Fe $(NO_3)_3$: Fe_2O_3) in a ratio of (B). 10:0, (C). 9:1, (D). 8:2, (E).7:3, (F).6:4, (G).5:5, (H). 4:6, (I). 3:7, (J). 2:8, (K). 1:9 and (L). 0:10.

3.1.3 FTIR spectra

The FTIR spectra of a M-CNFs performed with impregnation ratio of magnetic precursors of 10:0 to 0:10 has shown in figure 2. The dominant bands of the products at 525-575 cm^{-1} and 400-425 cm^{-1} are the characteristics of Fe_2O_3 [5]. The presence of C–O bond stretching of the C–O–C group at 1,096 – 1,121 cm^{-1}, respectively [6]. It can be seen that the FTIR spectra situated at 1387cm^{-1} revealed the characteristic of $Fe(NO_3)_3$, where this band was converted due to the carbonization process resulting the $Fe(NO_3)_3$ to change into Fe_2O_3 in Peak 400-425 cm^{-1} and 525-575 cm^{-1}.

3.1.4 XRD patterns

The pattern of the diffraction peak of M-CNFs at the position theta (theta; θ) in figure 3 shows the positions of the peaks in the 26°and 43° of MCNFs that are characteristic of crystalline graphite [7]. The peaks in these positions are low and wide in the range of 20-30°and 40-50°, respectively. Amorphous carbon could be explained between the heating process of the arrangement and it is the graphite form, and during carbonization there is a distribution of graphite structure. Higher carbonization heating results in smaller crystals of graphite [8]. However, the peak located at 27°, 32°, 37°, 45° correspond to Fe_2O_3, which appears in each spectra of M-CNFs and there are any peaks of $Fe(NO_3)_3$, therefore, it can be implied that the carbonization process causing the development of Fe $(NO_3)_3$ into Fe_2O_3 [18].

3.1.5 VSM analysis of M-CNFs

The magnetic properties of prepared M-CNFs was determined using a vibrating sample magnetometer is displayed in figure 4. M-CNFs was induced in magnetic field to collect the voltage signal which shows the relationship between magnetic moment and applied field [7]. It can be seen that M-CNFs performed at the impregnation ratio of 5:5 magnetic precursors (Fe_2O_3 and (Fe $(NO_3)_3$) has a magnetic property approximately 0.2 emu.

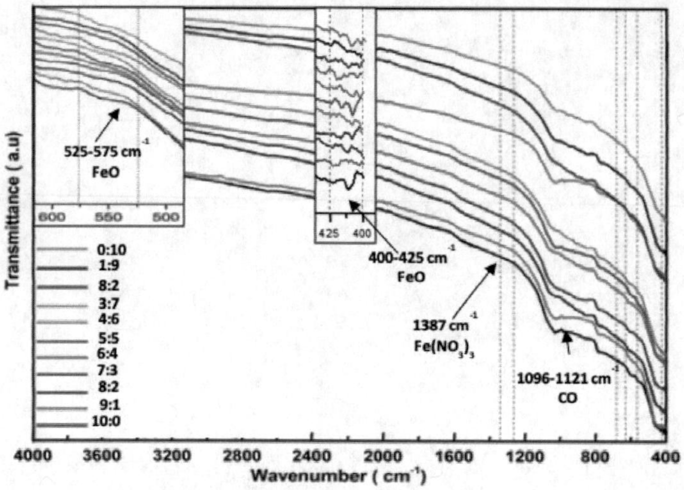

Figure 2. FTIR spectra of a MCNFs and add the iron nitrate and iron oxide (Fe $(NO_3)_3$: Fe_2O_3) in a ratio of (10:0) to (0:10).

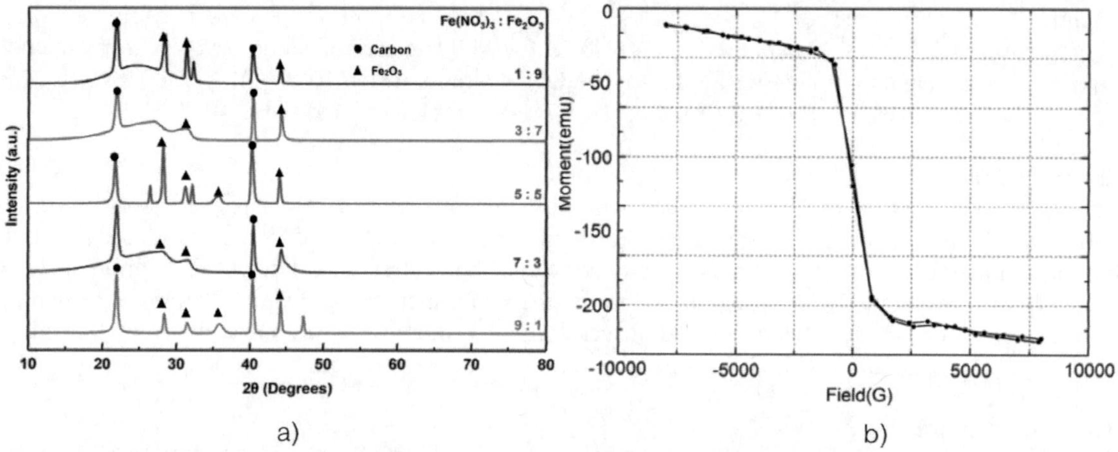

Figure 3. (a) XRD curves of the MCNFs and add the iron nitrate and iron oxide (Fe $(NO_3)_3$: Fe_2O_3) in a ratio of (9:1), (7:3), (5:5), (3:7), (1:9) and (b) VSM curve of M-CNFs performed at magnetic precursor impregnation ratio of 5:5.

3.2 Methylene blue adsorption

The results on methylene blue adsorption using as-prepared M-CNFs is shown on figure 4(a). It can be seen that the adsorption curves is linear relationship shows the gradual adsorption at the beginning until 5 min of adsorption time. That is because the adsorption on the pores is occurred rapidly due to the fresh active site on surface of M-CNFs leads to high adsorption rate. After 5 min, the adsorption is slightly decreased until equilibrium at the adsorption time of 15 min. The adsorption performance of M-CNFs is over 98 %. Nonetheless, M-CNFs can be separated conveniently by using a magnet as seen in figure 4(b).

IOP Conf. Series: Materials Science and Engineering **540** (2019) 012006 doi:10.1088/1757-899X/540/1/012006

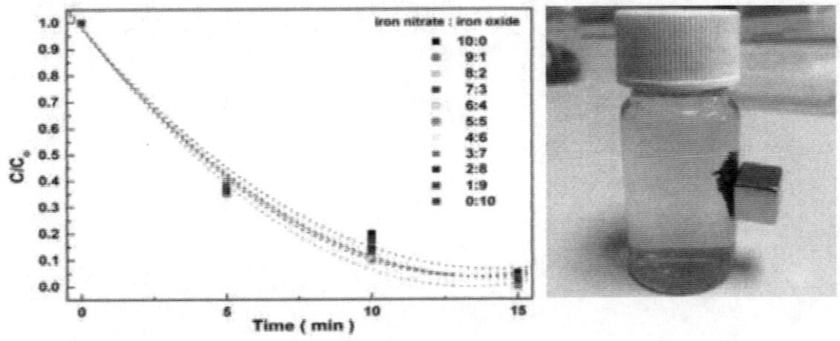

a b

Figure 4. (a) Adsorption capacity of M-CNFs on MB solution at the concentration of 100 mg/L and (b) separation of M-CNFs using a magnetic property.

4. Conclusions

In this study, the carbon nanofibers with magnetic feature (M-CNFs) have been prepared from horse manure via employing the optimum condition of hydrothermal treatment and carbonization processes. The resulting M-CNFs shows the magnetic property. This study reveal that the horse manure is perfectly alternative precursor to develop a carbon nanofiber because of its nature which can provide effective production cost and eco-friendly. Nonetheless, the M-CNFs can be applied as favourable adsorbent providing some prominent characteristics such as high carbon content and high specific surface area. The prepared M-CNFs with the impregnation ratio of 5:5 magnetic precursor shows a paramagnetic properties of 0.2 emu and the methylene blue adsorption is above 98%. It could be implied that animal manures is one of alternative precursors for a carbon-based adsorbent manufacturing.

5. Acknowledgment

The authors are thankful to the College of Nanotechnology, King Mongkut's Institute of Technology Ladkrabang and Department of Materials Engineering, Faculty of Engineering, Kasetsart University for facility support.

6. References

[1] Rajeev R and Amit K J 2016 Bioresource Technology **199** 92–102.
[2] Nathan M, Charles W, Bruce D, Richard E, Lee Y Y, Mark H and Michael L 2005 Bioresource Technology **96** 673–686.
[3] Toufiq M R, Joan G L, Helal U and Charles J C 2013 biomass and bioenergy **49** 86-94.
[4] Wei-Hong Y, Pei-Gao D, Feng W, Yu-Ping X 2016 Fuel **185** 229–35.
[5] Barron V, Torrent J 1996 J Colloid Interf Sci **177** 407–10.
[6] Ma X F, Chang P R, Yang J W and Yu J G 2009 CarbohydrPolym **75** 472–8.
[7] Suvarna M, Soumen D, Kalyan M and Subhadra C 2007 Nanotechnology **18** 275–84.
[8] Angin D, Köse T E and Selengil U 2013 Applied Surface Science **280** 705–10.
[9] Raisa K and Valerii B 2016 J. of Supercritical Fluids **117** 64–71.
[10] Bhandari P N, Kumar A, Bellmer D D and Huhnke R 2014 Renewable Energy **66** 346–53.
[11] Chao H, Apostolos G and Jing-Yuan W 2013 Applied Energy **111** 257–66.
[12] Chulhwan P, Myunggu L, Byunghwan L, Seung-Wook K, Howard A C, Jinwon L and Sangyong K 2017 Biochemical Engineering Journal **36** 59–65.
[13] Jun H, Dekui S, Shiliang W, Huiyan Z and Rui X 2014 Journal of Analytical and Applied Pyrolysis **106** 118– 24.
[14] Chao H, Apostolos G and Jing-Yuan W 2013 Applied Energy **111** 257–66.
[15] Daegi K, Kwanyong L and Ki Y P 2014 Fuel **130** 120–25.
[16] Jun H, Dekui S, Shiliang W, Huiyan Z and Rui X 2014 Journal of Analytical and Applied Pyrolysis **106** 118–24.

IOP Conf. Series: Materials Science and Engineering **540** (2019) 012006 doi:10.1088/1757-899X/540/1/012006

[17] Belver C, Munoz M A B and Vicente M A 2002 Chem. Mater. **14** 2033–43.
[18] Shaobin W A 2008 Dyes and Pigments **76** 714–20.

IOP Conf. Series: Materials Science and Engineering **540** (2019) 012007 doi:10.1088/1757-899X/540/1/012007

Effect of Water and Acid Pre-treatment on the Physicochemical Properties of Rice Husk for Silica Extraction

J T Librea[1], F D Dacanay[2], Z Z Martin[2], and L L Diaz[1]

1 Department of Mining, Metallurgical and Materials Engineering, College of Engineering, University of the Philippines, 1101, Diliman, Quezon City, Philippines
2 Materials Science and Engineering Program, College of Science, University of the Philippines, 1101, Diliman, Quezon City, Philippines
E-mail: jtlibrea@up.edu.ph

Abstract. Rice husks are residue materials from rice milling. These materials are mostly discarded after obtaining the rice, however, numerous studies have found out that rice husks are an excellent renewable source of silica. Rice husks are generally turned into rice husk ash in order to extract the silica in the material. Prior to the combustion process, pre-treatment methods are usually done in order to minimize the impurities in the husks. Failure to remove these impurities lead to undesired properties of the extracted silica. As such, we investigate some common pre-treatment methods, namely water and acid (HCl) treatment, and their effects on the surface morphology and elemental composition of the husks after treatment. It was found that H_2O was more effective in removing the bromide ions Br^-, while HCl becomes less effective at increasing concentration. Also, acid treatment was more effective in removing the sodium and potassium components compared to water treatment. However, efficiency decreases as the acid concentration increases due to interference of silica polymerization. Thus, for the removal of impurities in rice husk, lower concentration of acid pre-treatment is recommended.

1. Introduction

Silica is the most abundant material on the earth's crust. While crystalline silica has limited uses, amorphous silica has been found to be useful as fillers, absorbents and catalyst supports [1]. Silica nanoparticles have also been explored for medical use, such as drug delivery systems and biosensors [2], [3]. The increasing demands for such materials prompt the exploration of renewable sources for silica nanoparticles, one of which is rice husk ash [4]. Rice husk ash, which can be produced via controlled rice husk combustion, is found to have high silica content (around 95 %) [5]. However, several studies also stated that presence of impurities in the rice husk affect the quality of the rice husk ash [6], [7].

Impurities such as potassium and sodium oxides affect the surface of silica particles, decreasing the surface area, reactivity, and silica crystallization temperature [6]. To remove the impurities, acid leaching was facilitated [6], [7]. Leaching is an extraction of soluble constituent away from an insoluble solid. The process may cause physical and chemical changes because aside from removal of chemical impurities, leaching may also cause surface fragmentation due to the removal of the organic layer [8].

This work aims to compare the differences in the physical and chemical properties between untreated, water leached, and acid leached rice husks. The surface morphology of the resulting husks was investigated by scanning electron microscopy. Furthermore, elemental composition was analysed

IOP Conf. Series: Materials Science and Engineering **540** (2019) 012007 doi:10.1088/1757-899X/540/1/012007

through energy-dispersive x-ray spectroscopy and x-ray fluorescence spectroscopy. Overall, the efficacy of removing the impurities in both pre-treatment methods were investigated.

2. Methodology

2.1. Rice Husk Preparation and Pre-treatment
Rice husk (RH) samples were obtained from local farmers in Laguna, Philippines. The rice husks were washed with water to remove dirt and other impurities. The sample was air-dried. The treatments were comprised of control (non-leached), deionized H_2O, 0.5 M HCl, 1.0 M HCl, and 2.0 M HCl. Dried RH were soaked in the leaching agents for 48 hrs. The RH were then filtered, washed and neutralized (pH 7.0) with deionized water. The husks were then oven dried.

2.2. Determination of Surface Morphology
Scanning electron microscopy (SEM) imaging was done to determine the changes in surface morphology of the husk for untreated and different treatment methods. Rice husk was treated with ethanol to remove adsorbed water and oven-dried prior to SEM imaging [9]. The dried sample was then physically deposited onto a carbon tape and then it was coated with platinum for 30 secs using ion sputter. The sample was subjected under scanning electron microscopy (SEM) under 500 x and 1000 x magnifications at 5.0 kV accelerating voltage. The SEM equipment used for this study was Hitachi SU-8230 FE-SEM of the Department of Mining, Metallurgical, and Materials Engineering, College of Engineering, University of the Philippines, Diliman, Quezon City.

2.3. Compositional Analysis
Energy-dispersive x-ray spectroscopy (EDS) was done to determine percent composition for the underside and outer layer of the treated and untreated rice husks. It was simultaneously conducted with scanning electron microscopy at 205x magnification and 15.0 kV accelerating voltage.
X-ray fluorescence (XRF) was also done to determine percent composition for treated and untreated rice husks and consequently, it was used to determine the percent removal after different treatments. The XRF gun of HDXRF-XOS at the Extractive Metallurgy Lab II of the Department of Mining, Metallurgical, and Materials Engineering was the one used for this test.

3. Results and Discussions

3.1. Variation of Surface Morphology of Rice Husk with Treatment
Untreated and treated rice husks were subjected to SEM imaging to determine the changes in surface morphology after leaching (Figure 1). Smoothening was observed on the surface of RH after water and HCl leaching (Figures 1a, 1b, 1c). On the control sample, circular planar domes were observed. Then, with H_2O and HCl treatment, the domes were eroded and ridges were exposed. These ridges correspond with the silica-rich portion of the rice husk outer surface [10]. The ridges were sharper in H_2O samples compared to HCl. This may be caused by the removal of organic layer on the outer shell of the rice husk.
SEM imaging was also performed on the RH underside, which is mostly composed of fibres. The control underside had homogeneous flat structure that showed the distinctive feature of the organic fibre layer of the rice husk [8], [10]. Etching was observed on the surface of the treated samples relative to the untreated sample (Figures 1d, 1e, 1f). Water treated samples had fibre-like structures with minimal distinct features. On other hand, acid treated samples showed individual fibres that confirm removal of organic layer on the underside of the rice husk samples. Elemental analyses were then done to determine the composition of the removed material.

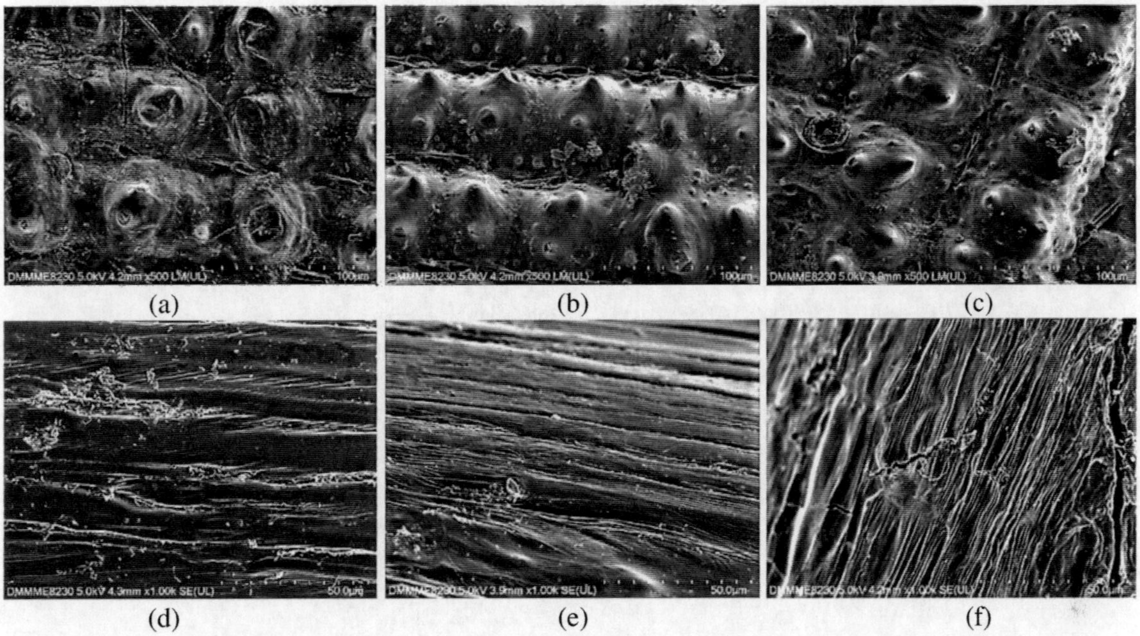

Figure 1: SEM images of (1a) untreated, (1b) H_2O, and (1c) 2.0 M HCl treated rice husk lemma. SEM images of the rice husk underside for (1d) untreated, (1e) H_2O, and (1f) 2.0 M HCl treated are also shown.

3.2. Elemental Composition

Energy-dispersive x-ray spectroscopy (EDS) was done on the outer and underside part of the husks. For the outer shell of RH, it was observed that there was an increase in C percentage, but Mg and Al were undetectable after being leached with water and HCl (Table 1). Furthermore, K and Ca were completely undetected after leaching with 2.0 M HCl. The increase in the carbon content supports the SEM image that showed exposure of the RH shell.

Table 1. Energy-Dispersive X-Ray Spectroscopy Results (Outer shell)

% Weight	C	O	Mg	Al	Si	K	Ca	Cl
Control	26.48	51.70	0.13	0.75	20.37	0.20	0.36	—
H_2O	31.97	48.29	—	—	19.55	0.09	0.09	—
2.0 M HCl	30.96	48.47	—	—	20.57	—	—	—

For the underside of the RH, there was an observed decrease in C content after treating with H_2O and 2.0 M HCl (Table 2). Mg, Al, and K were also undetectable after treatment with water and acid. However, Cl was detected in acid treated sample that may be due to trapped acid in the rice husk fibres. After leaching with water and acid, the silica content in the underside portion was observed to increase. This may be due to the removal of the organic layer as observed in the SEM image (Figure 1).

Table 2. Energy-Dispersive X-Ray Spectroscopy Results (Underside)

% Weight	C	O	Mg	Al	Si	K	Ca	Cl
Control	45.72	46.03	0.06	—	7.98	—	0.20	—
H_2O	30.80	48.61	—	—	20.43	—	0.16	—
2.0 M HCl	28.07	46.23	—	—	25.35	—	0.14	0.20

Based on the XRF data (Table 3), it was found that most of the outer surface material were impurity elements, i.e. Mg, Al, K, and Ca. After washing with water and HCl, a decrease in the concentration of these elements was observed, which led to the smoothening of the surface morphology (Figure 1).

Table 3. X-Ray Fluorescence Spectroscopy Results

Elements	Br	Ca	Fe	K	Si	Sr	Mn	Rb	Zn	Ti
Control	0.007	0.520	0.097	0.771	98.451	0.002	0.004	0.001	0.001	—
H_2O	—	0.608	0.090	0.135	98.945	0.016	0.182	—	0.018	—
0.5 M HCl	—	0.105	0.048	0.050	99.788	0.009	—	—	—	—
1.0 M HCl	0.002	0.171	0.043	0.028	99.748	0.002	0.004	0.001	0.001	—
2.0 M HCl	0.004	0.471	0.049	0.029	99.437	0.005	—	—	0.005	—

3.3. Removal of Impurities

X-ray fluorescence spectroscopy was performed on all systems to determine percent composition of the impurity elements (Table 3). The percent removal relative to the control was calculated by the formula,

$$\%_{removed} = \left(\frac{C_{control} - C_{treated}}{C_{control}}\right) \times 100$$

where $C_{control}$ is the concentration of the element for the untreated system, and $C_{treated}$ is the concentration of the element for a specific treatment.

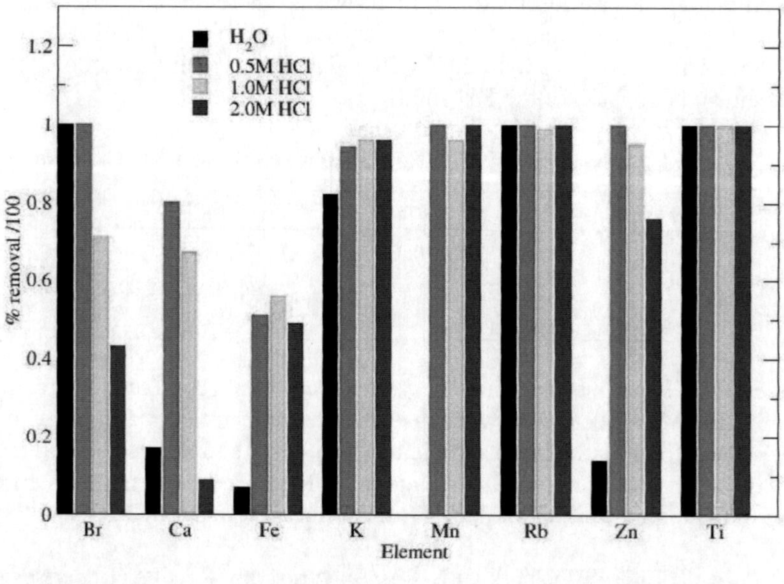

Figure 2. Percentage removal of impurities based on the XRF data.

It was observed that trace amounts of Rb and Ti were present in the control and completely removed after water and acid treatment (Table 3 & Figure 2). It was also observed that HCl was much more favourable in removing Fe, Mn, and Zn, which were also found at trace amounts in the control.

Water was found to be ineffective in removing some of the metals. In solution, water weakly ionizes to form H^+ and OH^-:

$$H_2O \rightleftharpoons H^+ + OH^- \qquad (1)$$

Free OH ions can react with the metal impurities Ca, Fe, Zn, and Mn to form the corresponding hydroxides [11] according to the following reactions [12], [13]:

$$Ca^{2+} + 2\,OH^- \rightleftharpoons Ca(OH)_2 \qquad (2)$$
$$Fe^{3+} + 3\,OH^- \rightleftharpoons Fe(OH)_3 \qquad (3)$$
$$Zn^{2+} + 2\,OH^- \rightleftharpoons Zn(OH)_2 \qquad (4)$$
$$Mn^{x+} + x\,OH^- \rightleftharpoons Mn(OH)_x \qquad (5)$$

These weakly soluble hydroxides are not easily removable by water. Mn can be present as Mn^{2+}, Mn^{3+}, or Mn^{4+} [14], and any of these may form corresponding hydroxides which can undergo further oxidation. These may account for the low percent removal of the metal ions during water treatment.

In addition to removing Br, Rb, and Ti, 0.5 M HCl was found to be more effective in removing Ca, Fe, K, Mn, and Zn. The hydrogen ion is more reactive compared to the ionic forms of these metals (e.g. Ca^{2+}, Fe^{2+}, K^+, Mn^{2+}, and Zn^{2+}) [11] ~ [13]. Thus, H^+ can displace these ions in the rice husk matrix, enabling effective leaching of these metal impurities. However, efficiency decreases as the acid concentration increases. Gorrepati et al. [15] report that the rate of silica polymerization rises exponentially with increasing pH. The formation of silica polymers may interfere with the ion-exchange mechanism between H^+ and the metal ions, thus lowering the amount of metal impurities removed.

Potassium is found to inhibit carbon removal during the rice husk combustion [16]. Ca^{2+} and OH^- can damage the Si-O bonds of amorphous silica [17]. It is desired to remove as much of these metals as possible to ensure the production of high purity silica from rice husk. Thus, from our results, 0.5 M HCl is recommended for use in rice husk pre-treatment for silica production.

4. Conclusion

Pre-treatments were able to remove impurities in the rice husk. Water washing was found to be effective in removing Br, K, Rb, and Ti, but not Ca, Fe, Mn, and Zn. HCl was effective in removing Ca, Fe, Mn, and Zn, with a higher efficiency in removing Br, K, Rb, and Ti. However, efficiency decreases as acid concentration increases. Therefore, low HCl concentration is recommended for rice husk pre-treatment.

References

[1] Gurav J L, Jung I K, Park H H, Kang E S and Nadargi DY 2010 Silica aerogel: synthesis and applications *J. of Nanomat.* JNM(2010)409310
[2] Slowing I I, Vivero-Escoto J L, Wu C W and Lin S Y L 2008 *Adv. Drug Deli. Rev.* **60** 1278-1288
[3] Lu J, Liong M, Zink J I and Tamanoi F 2007 *Small J.* **38** 1341-1346
[4] Soltani N, Bahrami A, Pech-Canul M and González L 2015 *Chem. Eng. J.* **265** 899-935
[5] Rafiee E, Shahebrahim S, Feyzi M and Shaterzadeh M 2012 Optimization of synthesis and characterization of nanosilica produced from rice husk a common *Int. Nano Letters* **2** INL28(2008)5326
[6] Kurama H and Kurama S 2003 *18th Int. Mining Congress and Exhibition of Turkey* pp 431-435
[7] Chandrasekhar S, Pramada P N and Praveen L 2005 *J. of Mat. Sci.* **40** 6535-6544
[8] Patel M, Kakera A and Prasanna P 1987 *J. of Mat. Sci.* **22** pp 2457-2464
[9] Weiss C A and Moser R D 2015 *Environmental Consequences of Nanotechnologies* (The US Army Engineer Research and Development Center: Washington, DC.)
[10] Park B-D, Wi S, Lee K, Singh A P, Yoon T-H and Kim Y 2003 *Biomass and Bioenergy* **253** 318-327
[11] Brown T L, LeMay J H E, Bursten B E, Murphy C J and Woodward P M 2012 *Chemistry: The Central Science* 12th ed (Upper Saddle River: Pearson Prentice Hall)

[12] Skoog D A, West D M, Holler F J and Crouch S 2015 *Fundamentals of analytical chemistry* 9th ed (Belmont CA USA: Brooks/Cole1 Publishing Company)
[13] Harris D C 2015 *Quantitative chemical analysis* 9th ed (New York: WH Freeman and Company)
[14] Xiong L, Saito K, Sekiya E H, Sujarid-worakun P and Wada S 2009 *J. of Met., Mater. and Min.* **19** 73-77
[15] Gorrepati E A, Wongthahan P, Raha S and Fogler S 2010 *Langmuir* **26** 10467-10474
[16] Krishnarao R, Subrahmanyam J and Kumar T 2001 *J. of Euro. Cera. Soc.* **21** 99-104
[17] Yu Q, Sawayama K, Sugita S, Shoya M and Isojima Y 1999 *Cem. and Conc. Res.* **29** 37-43

Acknowledgments
The authors thank the Department of Mining, Metallurgical, and Materials Engineering for the use of the scanning electron microscopy (SEM), energy-dispersive X-ray spectroscopy (EDX), and X-ray fluorescence spectroscopy (XRF) facilities. The authors also thank the Engineering Research and Development for Technology of the Department of Science and Technology (DOST-ERDT) for the funding of the research.

IOP Conf. Series: Materials Science and Engineering **540** (2019) 012008 doi:10.1088/1757-899X/540/1/012008

Effect of Recycled Rubber-derived Carbon (CRR) on the Thermal Stability, Cure Behaviour, and Dynamic Mechanical Property of NR/SBR Compound

E C Danila[1], M Lanwang[2] and L L Diaz[1]

1 Department of Mining, Metallurgical and Materials Engineering, College of Engineering, University of the Philippines, 1101, Diliman, Quezon City, Philippines
2 Materials Science and Engineering Program, College of Science, University of the Philippines, 1101, Diliman, Quezon City, Philippines
E-mail: ecdanila@up.edu.ph

Abstract. From an economic viewpoint, locally-sourced carbon from recycled rubber was identified to be a potential filler substitute to commercial carbon black in rubber tire. In this research, methodical measures were used to compare several properties of rubber compounds reinforced with carbon from recycled rubber (CRR) and commercial carbon black fillers to systematically qualify such substitution. Thermogravimetric analysis elucidated the overall thermal stability of the compounds, the degradation behaviour of their components and the purity of the fillers used. Differential Scanning Calorimetry and Rubber Process Analyzer test results revealed the difference in polymer network contribution and filler filler interaction observed in the compounds, which collectively described the reinforcing effect of the fillers in the material. Conclusively, the difference in composition of fillers used resulted in the alteration of reinforcing effects in the rubber compounds.

1. Introduction

Rubber, as one of the most significant elastomeric composite material, exhibits remarkable viscoelastic properties that enable it to become an advantageous choice for various useful applications in the industry and everyday use. Such interesting properties arise from the integrated characteristics of its components which are the base elastomer, fillers, and additives [1], [2]. Of all its components, rubber fillers have been the focus of studies for decades. It has been proven that the physical, mechanical and thermal properties of the rubber compound can be enhanced by adding fillers [3], [4]. In the paper published by Fröhlich et. al., the primary filler factors affecting rubber reinforcement were thoroughly discussed using an elastic modulus curve. Such factors include the strain-dependent filler-filler interaction and the strain-independent polymer network contribution. The filler-filler interaction is highly dependent on the particle size, surface area, and amount of loading of the filler which, when treated as a whole, signifies the effective contact area between the filler and polymer matrix; while the polymer network contribution is highly related to the crosslink density of the matrix and the polymeric nature [5].

Since rubber has already been part of our daily lives, eco-friendly and sustainable materials for its production are also being widely studied. Aside from the traditional yet expensive filler, carbon black, many fillers from different sources are now being produced, studied, and reinforced with each other [3]. Now, as materials scientists and engineers, it is our roles to qualify and characterize these new fillers and to determine how they affect our rubber.

IOP Conf. Series: Materials Science and Engineering **540** (2019) 012008 doi:10.1088/1757-899X/540/1/012008

In the Philippines, the Rubber Project 3, a project under the umbrella of National Research and Development Program for Rubber Processing and Rubber Manufacturing, implemented through the University of the Philippines – Department of Mining, Metallurgical, and Materials Engineering (UP – DMMME), has identified several potential carbon black filler substitutes that can be sustainable and locally-sourced. One of the filler substitutes recognized was a pyrolytic carbon black derived from the pyrolysis of recycled rubber tires called Carbon from Recycled Rubber (CRR). CRR is a fine black powder with properties shown in table 1 [6].

Table 1. Comparison of properties of filler substitute CRR against the commercial high abrasion furnace (HAF) carbon black filler [6].

Properties	HAF	CRR
Major Elemental Impurities	S, K, Ca	Zn, S, Ca
Particle Size (80% passing)	x>500 μm	X<150 μm
Surface Area, m^2/g	58.11	32.14
Iodine Adsorption Number, g/kg	121.04	177.44
pH	7.15	8.51

In this research, the effect of CRR as filler substitute to the in terms of thermal and viscoelastic properties of the synthesized compounded rubber material was evaluated. Specifically, the effect of filler substitution on thermal stability, polymer network contribution, and filler-filler interaction were determined.

2. Materials and methods

2.1. Sample preparation

Rubber compounds were blended using a Bharaj BMMM-6 twin roll mill in the Rubber Processing Laboratory of UP-DMMME. A specific tire wirebead formulation for compounding shown in table 2 dictated the amount of each reagent used. Compounded rubber were vulcanized using a thermal press in the Rhodeco Rubber Processing Services, Incorporated. Vulcanized products were properly stored prior to analyses.

Table 2. Formulations of compounded rubber material filled with the commercial filler (HAF) and substitute filler (CRR). [6]

Compounding Ingredient	HAF-filled (phr)	CRR-filled (phr)
Elastomer	100	100
CRR	-	135
HAF	135	-
Additives	81.1	81.1

2.2. Thermogravimetric analysis

The thermogravimetric curves of rubber compounds were determined using ASTM D6370 - 99 [7] in a TA Instruments Q500 Thermogravimetric Analyzer in the Materials Characterization Laboratory of UP-DMMME. Curves were analyzed to identify the composition and degradation temperatures of the components in the compounds.

2.3. Differential scanning calorimetry

The heating profiles of unvulcanized rubber compounds were established following the procedure stated by ASTM D3418-12 [8] in a Q200 Differential Scanning Calorimeter in the D&L Industries to determine curing characteristics and polymer network contribution of both compounds.

2.4. Rubber process analysis

The elastic modulus curves of rubber compounds were established by employing ASTM 6601 [9] in a Rubber Process Analyzer in the Mechanical Testing Laboratory of UP-DMMME to investigate the filler-filler networking within the compounds.

3. Results and discussion

3.1. Compositional Analysis and Thermal Stability Measurements

Figure 1 shows the thermogravimetric (TG) curve, in green, and the derivative (DTG) curve in blue, of the HAF compound.

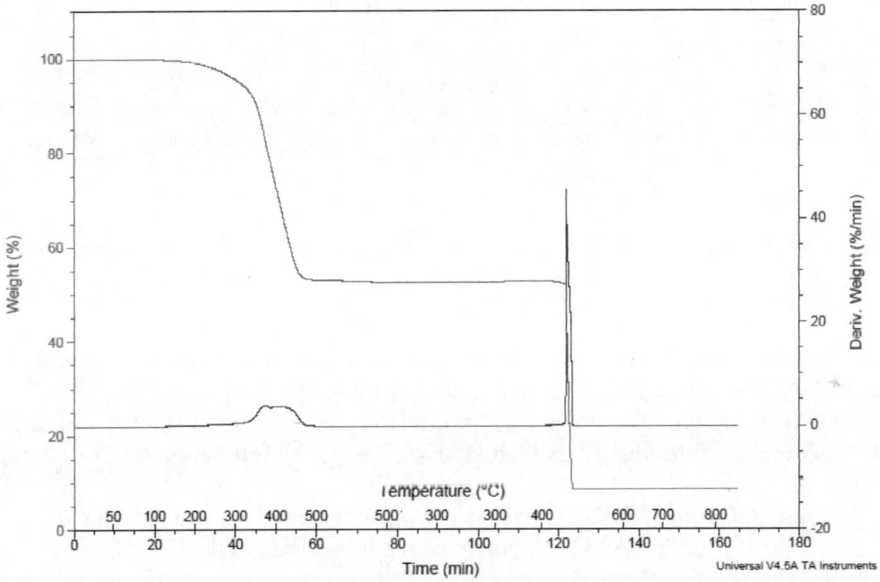

Figure 1. TG and DTG plots showing the two-stage degradation and the three degradation events of the HAF compound.

The TG curve showed the two-stage degradation of the HAF compound. The first shoulder signifies the degradation of elastomers while the second is that of carbon. From the DTG, 2 peaks were observed, a broad peak was seen around 350°C to 500°C and a sharp peak at 590.44°C. A curve fitting software was used to accurately identify the events that occurred in the broad region. Two identified peaks at 388.71°C and 442.83°C were elucidated to be the degradation temperature of natural rubber (NR) and styrene-butadiene rubber (SBR) [10], while the peak at 590.44°C confirmed the degradation of carbon [7]. Analyzing the TG, the composition of the compound was determined following the calculations stated from ASTM D6370 – 99. It was found out that the HAF compound contained 47.42% volatile organic compounds (VOC), 44.16% carbon, and 8.45% of remaining ash.

For the CRR compound, analysis of its TG and DTG curve, shown in figure 2, gave rise to its composition and the degradation temperature of its components. Similarly, the TG and DTG curve also showed two-stage degradation, and three degradation events. From the curves, it was found out that the compound contained 47.89% VOC, 37.51% carbon, and 14.54% of remaining ash and the degradation peaks of its components were at 385.15°C, 435.07°C, and 488.13°C.

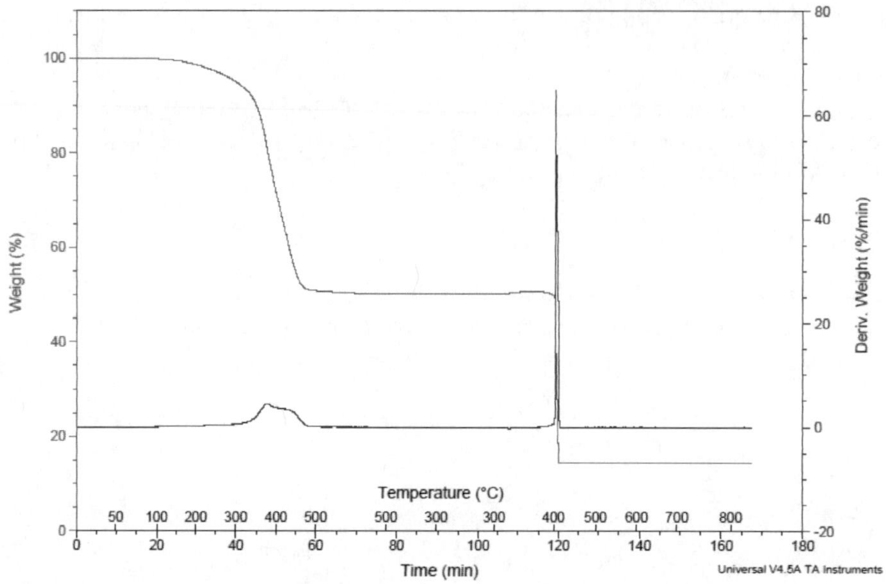

Figure 2. TG and DTG plots showing the two-stage degradation and the three degradation events of the CRR compound.

Calculated compositions and degradation temperatures were compared and are presented in tables 3 and 4. Calculations showed that the HAF compound contained more carbon and less VOC and ash content as compared to the CRR compound. These differences can be attributed to the purity of the fillers used. CRR contains more impurities than HAF that might be remnants from its source material.

Table 3. Compositional analysis showing substantial differences in the carbon and ash content of the two compounds.

Component	HAF-filled (phr)	CRR-filled (phr)
Volatile Organic Compounds, %	47.42	49.79
Carbon Content, %	44.16	35.71
Ash Content, %	8.425	14.54

Comparing, identified degradation peak values of CRR occurred in lower temperature in contrast to the peaks found in the HAF-filled compound. The most noticeable difference was the degradation temperature of carbon in CRR-filled compound at 488.13 °C, which is a 100°C earlier than its HAF-filled counterpart. This degradation temperature confirms that the carbon in CRR is a pyrolytic carbon black as expected based on the thermal history that its source has undergone [4].

Table 4. Lowered degradation temperatures were observed for CRR-filled rubber compounds.

Component	HAF-filled	CRR-filled

	(°C)	(°C)
Natural Rubber	388.71	385.15
Styrene-Butadiene Rubber	442.83	435.07
Carbon	590.44	488.13

3.2. Polymer network contribution

Figure 3 shows vulcanization peaks snipped from the DSC heating profile of the unvulcanized rubber compounds. It is observed that the vulcanization peak of the CRR-filled compound is less intense than that of the reference HAF-filled compound.

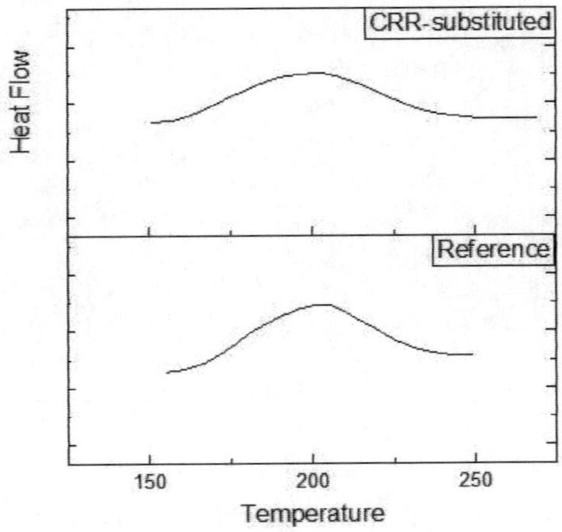

Figure 3. The extent of curing is reduced when CRR is used as filler as manifested by the decrease in peak height.

In table 5, it is shown that the substitution of HAF with CRR lowers the curing temperature and the vulcanization peak values. The impurities found on the high ash content-CRR might have obstructed the cross-linking process which resulted in the inferior cure properties of CRR-filled rubber compounds.

Table 5. Lowered cure characteristics were observed for CRR-filled rubber compounds.

	HAF-filled	CRR-filled
Curing Temperature, (°C)	201.50	198.88
Peak Height	0.0773	0.0616
Peak Area	3.2238	3.0035

This decrease in the heat of curing indicates fewer cross-links formed in the CRR-filled compound after vulcanization, which in turn, lowers the polymer network contribution.

3.3. Filler-filler interaction

The strain amplitude dependence of complex modulus of the rubber compounds is presented in figure 4. Despite CRR having smaller particle size than HAF, lowered G' values throughout the strain sweep were still observed, deviating from the expected results. In theory, CRR having more effective contact area should have a higher reinforcing effect than HAF filler [5]. The researchers believe that this deviation from theory is a consequence of the impurities in the CRR filler, in agreement to the results of our thermogravimetric and cure characteristic analysis. The substantial amount of non-reinforcing impurities present, i.e. the increased ash content, played a major role in lowering the filler-filler interaction in the compound. The reinforcing effect of carbon present in CRR is hindered by the non-reinforcing effect of impurities it contained.

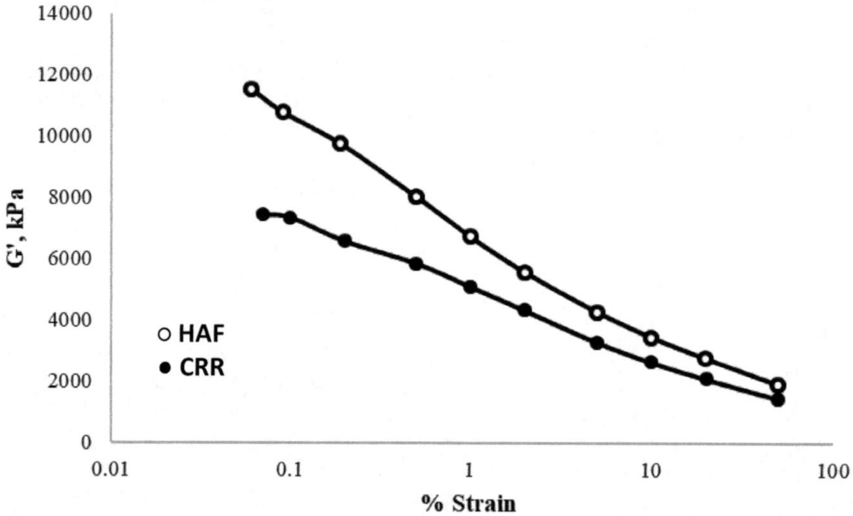

Figure 4. Lowered strain amplitude dependence of complex modulus was observed for the CRR-filled compound.

4. Conclusions

Thermogravimetric analysis of rubber compounds revealed that the use of CRR instead of HAF as fillers lowers the overall thermal stability of a rubber material as the carbon component of the CRR-filled compound degrades at a much lower temperature than that of the HAF-filled compound. DSC and RPA tests expose the lowered reinforcing effect when CRR filler is used due to the decrease in polymer network contribution and filler-filler interaction. CRR-filled compound showing inferior observed properties as compared to the HAF-filled compound is a result of the substantial amount of non-reinforcing impurities present in the CRR filler as observed in the compositional analysis. Researchers recommend the employment of beneficiation techniques to CRR to further remove non-reinforcing impurities present.

5. References

[1] Wang J and Chen D 2013 *J. Nanomater.* **2013** 496584
[2] Leblanc J L 2002 *Prog. Polym. Sci.* **27** 627–687
[3] Polynuclear aromatic hydrocarbons, Part 2, Carbon blacks, mineral oils (lubricant base oils and derived products) and some nitroarenes 1984 *IARC Monogr. Eval. Carcinog. Risk. Chem. Hum.* **33** 1–222
[4] Berki P, Göbl R and Karger-Kocsis J 2017 *J. Polym. Test.* **61** 404-415
[5] Fröhlich J, Niedermeier W and Luginsland H 2005 *Compos. Appl. Sci. Manuf.* A **36(4)** 449-460.
[6] Diaz L 2017 *Enhancing and increasing local content in rubber for motorcycle tire application* Quezon City, Department of Mining, Metallurgical and Materials
Engineering, University of the Philippines
[7] ASTM Standard D6370 – 99 2003 *Standard Test Method for Rubber – Compositional Analysis by Thermogravimetry (TGA)* ASTM International, West Conshohocken, PA 19428-2959 United States

[8] ASTM Standard D3418 2015 *Standard Test Method for Transition Temperatures and Enthalpies of Fusion and Crystallization of Polymers by Differential Scanning Calorimetry* ASTM International, West Conshohocken, PA 19428-2959 United States

[9] ASTM Standard 6601 2015 *Standard Test Method for Rubber Properties—Measurement of Cure and After-Cure Dynamic Properties Using a Rotorless Shear Rheometer* ASTM International, West Conshohocken, PA 19428-2959 United States

[10] Varkey J, Augustine S and Thomas S 2010 *Polym. Plast. Technol. Eng*. **39(3)** 415-435

Acknowledgments

The authors would like to thank the Department of Mining, Metallurgical, and Materials Engineering for the use of their twin roll mill, thermogravimetric analyzer, and rubber process analyzer in the Materials R&D and Consulting Facility laboratories. The authors also thank the Engineering Research and Development for Technology of the Department of Science and Technology (DOST-ERDT) for the funding of this research.

IOP Conf. Series: Materials Science and Engineering **540** (2019) 012009 doi:10.1088/1757-899X/540/1/012009

Transmissibility, Isolation Region and Damping Characteristics of Gray Cast Iron and Aluminium Composite

T.G.T. Nindhia[1]

1 Mechanical Engineering, Engineering Faculty, Udayana University, Jimbaran, Bali, Indonesia
E-mail: nindhia@yahoo.com/tirta.nindhia@me.unud.ac.id

Abstract. All The metal composite was prepared in this research by utilizing waste of aluminum and gray cast iron chips in supporting the global program of sustainable recycling-oriented society. The process was carried out via hot extrusion at 400°C. Three compositions were provided in this research, namely the composition of aluminum chips with addition 0 vol.%, 20 vol.%, and 40 vol.% of gray cast iron chips. Increasing the gray cast iron was found affect on reducing the natural frequency of the product. Addition about 20 vol.% of gray cast iron decreases the value of damping capacity. Conversely, addition about 40 vol.% gray cast iron increases the value of damping capacity significantly. It was also found that the addition of gray cast iron could extend the isolation region and decreased the transmissibility at the isolation region.

1. Introduction

These The material of aluminum and the gray cast iron are widely used, so that also produce a large scale of waste from the manufacturing processes. In relation with supporting the global program of sustainable recycling-oriented society, the waste of aluminum and gray cast iron chips were utilized in this research. The future application of this composite is addressed for functional and useful material such as passive vibration isolator. Gray cast iron was well known as a material that having a high damping capacity, due to flake-like shape of graphite that was contain in its microstructure and widely used as passive vibration isolator [1].

Transmission of unwanted vibration can cause various problems in sensitive system including system malfunction and performance degradation. The common approach to limit the transmission of harmful vibration in the sensitive system is by adapting passive vibration isolator[2]. Since the mechanism of passive vibration isolator was depend on the damping property of the material, it is hoping that the use of gray cast iron chips for passive vibration isolation will come in to a positive result.

The aluminum chips was used asamatrix since known having wide range of plasticity, so that possible to be processed via hot extrusion. Hot extrusion was introduced in this research since previous research proved that this method resulting high damping product [3-4].

2. Experimental

The waste of gray cast iron chips from the type of FC 150 (Compositions in Table 1) was obtained directly from the company of Aishin Takaoka, Japan. The simulation of waste aluminum chips was made by using milling machine with cutting condition at 1005 rpm, feed at 1000 mm/min., with deep of cut at 0.1 mm. The aluminum was from the type of Al 6061-T6 (Composition in Table 2).

Table1. Chemical composition of FC 150.

C	Si	P	S	Mn	Fe
3.1	1.6	0.07	0.05	0.28-1.01	Balance

Table 2. Chemical composition of Al 6061-T6.

Si	Fe	Cu	Mn	Mg	Cr	Zn	Ti	Others	Al
0.40-0.8	≤0.7	0.15-0.40	≤0.15	0.8-1.2	0.04-0..35	≤0.25	≤0.15	≤0.15	Balance

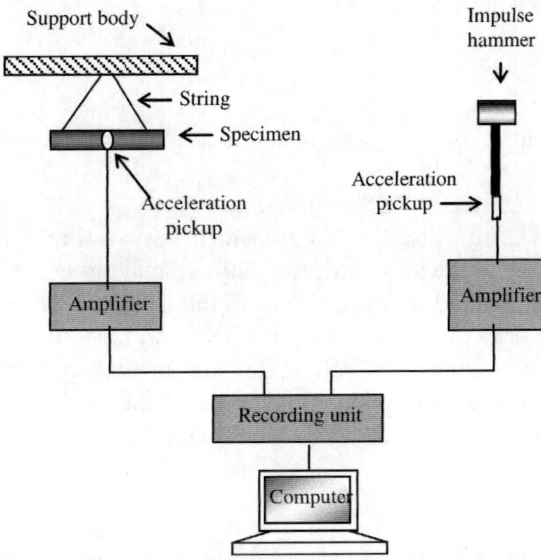

Figure 1. Out line chart of experimentalapparatus.

The chips were cleaned ultrasonically in ethanol, afterward mixed and homogenized in V-type tube mixer. Three compositions were provided in this research namely, composition of Al chips with addition 0, 20, and 40 vol.% of gray cast iron chips. The Process was continued with compacting at 1 MPa to become a billet with 70 mm length and 60 mm in diameter. The billets were then covered with 0.01 mm aluminum foil for hot extrusion preparation. The hot extrusion was carried out at 400°C, with applying pressure at 500 Ton, the speed was arranged at 73 mm/min. The extrusion ratio was 4. The lubricant that was used at the die was MoS_2, mean while the graphite colloid was utilized as lubricant at main cylinder of extrusion. Final diameter of the product was 15 mm.

The damping capacity of the product, natural frequency, and transmissibility were measured by utilizing hammering vibration test. figure 1 shows the outline chart of experimental apparatus. The specimen with diameter 15 mm and 150 mm length was hung, and then hit by using impulse hammer (GK-300, ONO SOOKI). The acceleration pickup sensor at the specimen and hammer transmitted the vibration signal to the amplifier to be forwarded to the recording unit (LX-10, TEAC). The signal data was then extracted to the computer to be analyzed its natural frequency, damping capacity, transmissibility, and the range of isolation region.

3. Results and discussion

The The sampling frequency (fs) was set at 48.000 Hz, with total number of discrete samples (N) that was taken is 4096. The result of damping signal can be seen at figure. 2. The value of damping capacity was expressed by attenuation coefficient (α) and was obtained by fitting the graph to the equation:

$$A(t) = A_o \exp[-\alpha t] \quad (1)$$

A(t) was the amplitude at any time, and initial amplitude was expressed by A_o. As can be seen at figure. 2, the value of α decreased from 25.3[S^{-1}] to become 16[S^{-1}] by adding of 20% gray cast iron, and by adding about 40% gray cast iron the value of α increased surprisingly to become 43[S^{-1}]. As a comparison [5], magnesium was known as a single metal that have high value of damping capacity, and its α value was 30.5[S^{-1}].

To obtain the natural frequency of the product, the time domain at figure 2 should transform to become frequency domain by using a discrete Fourier transform:

$$F(k\Delta f) = \sum_{n=o}^{N-1} f(n\Delta t)e^{-i(2\pi k\Delta f)(n\Delta t)} \quad (2)$$

for k= 0, 1, 2, …, N-1

The time increment between samples wasΔt, and frequency increment wasΔf. The result was a complex number z = x + iy. The magnitude of z was given by $|z| = \sqrt{x^2 + y^2}$.

For faster calculation of equation 2, the fast Fourier transforms was utilized in this research. After frequency domain was obtained, the frequency with has highest magnitude at 1st Nyquist zone was mentioned as a natural frequency (f_r) of the product. It was found that the value of natural frequency of the specimen without addition of gray cast iron was 2613.3 Hz.Addition of 20 vol.% gray cast iron make the natural frequency decrease to become 2191.4 Hz., and increasing the gray cast iron content up to 40 vol.% make the value of natural frequency decrease until 2027.3 Hz as can be seen in figure.3.

The value of logarithmic decrement (δ) can be obtained from the equation:

$$\alpha = \pi f_r Q^{-1}$$

$$Q^{-1} = \frac{\delta}{\pi} \quad (3)$$

$Q^{-1}$was an internal friction.

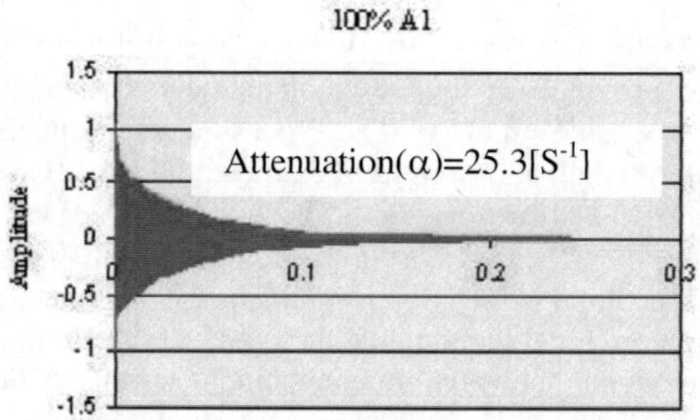

(a)

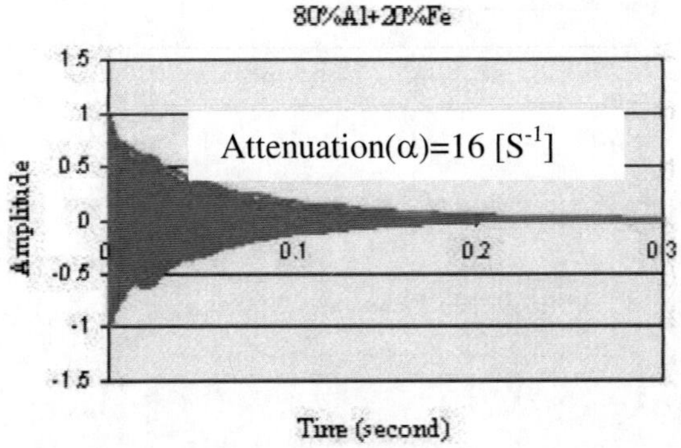

(b)

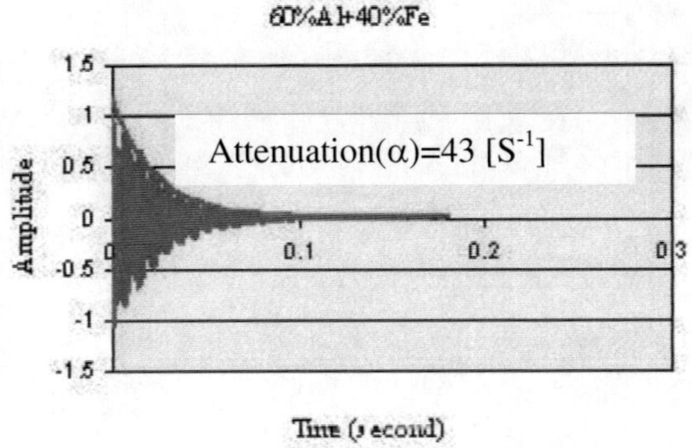

(c)

Figure 2. The damping signal of the product and its damping value that was expressed in attenuation coefficient (α). a.100% Al. b. 80% Al+20%Fe, c. 60% Al+40%Fe

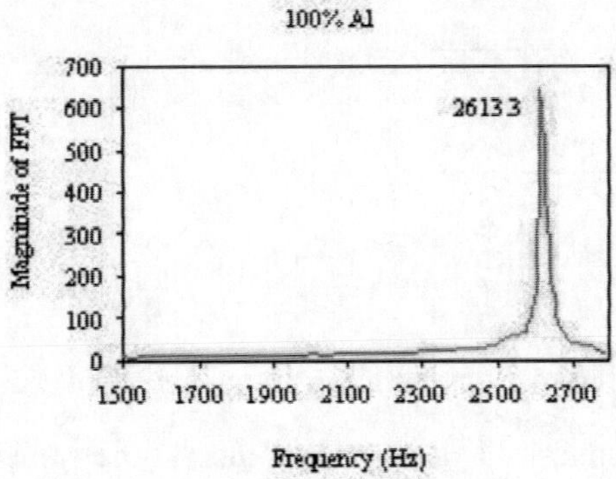

(a)

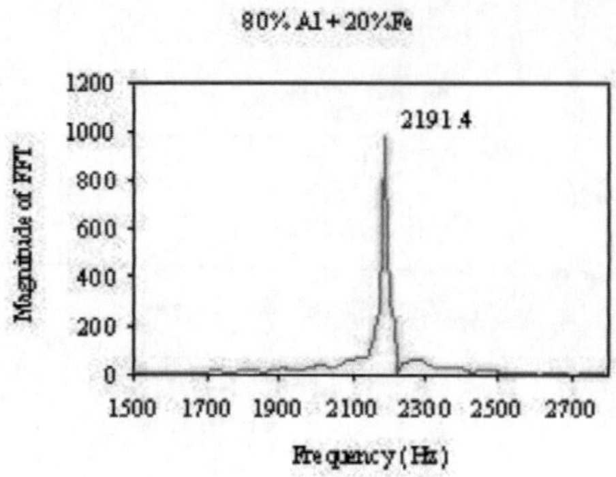

(b)

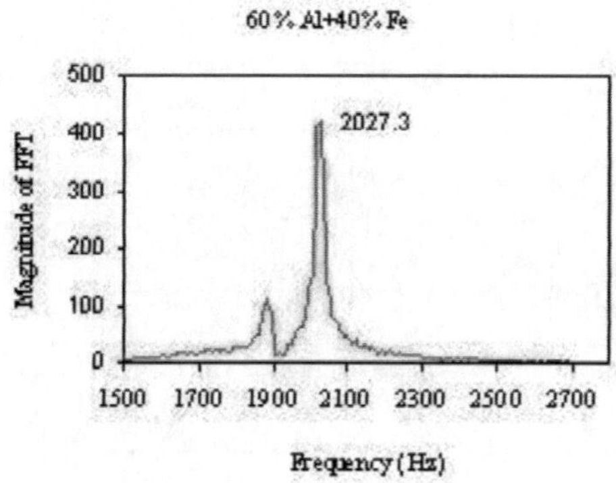

(c)

Figure. 3. The value of natural frequency that was obtained from the Fourier transform. a. 100% Al, b. 80%Al+20%Fe, c. 60%Al+40% Fe.

The value of logarithmic decrement that was obtained from Equation 3 and the value of natural frequency taken from figure. 3 were used to obtain the transmissibility (μ) that was obtained by using Equation 4 suggested by [6,7].

$$\mu = \sqrt{\frac{1+\left(\frac{\delta}{\pi}\right)^2}{\left[1-\left(\frac{f}{f_r}\right)^2\right]^2+\left(\frac{\delta}{\pi}\right)^2}} \qquad (4)$$

Where f represent frequency of interest or operating frequency. Isolation region was a region which the value of μ bellow 1. The transmissibility will have value 1 if the value of $\frac{f}{f_r}$ is $\sqrt{2}$.

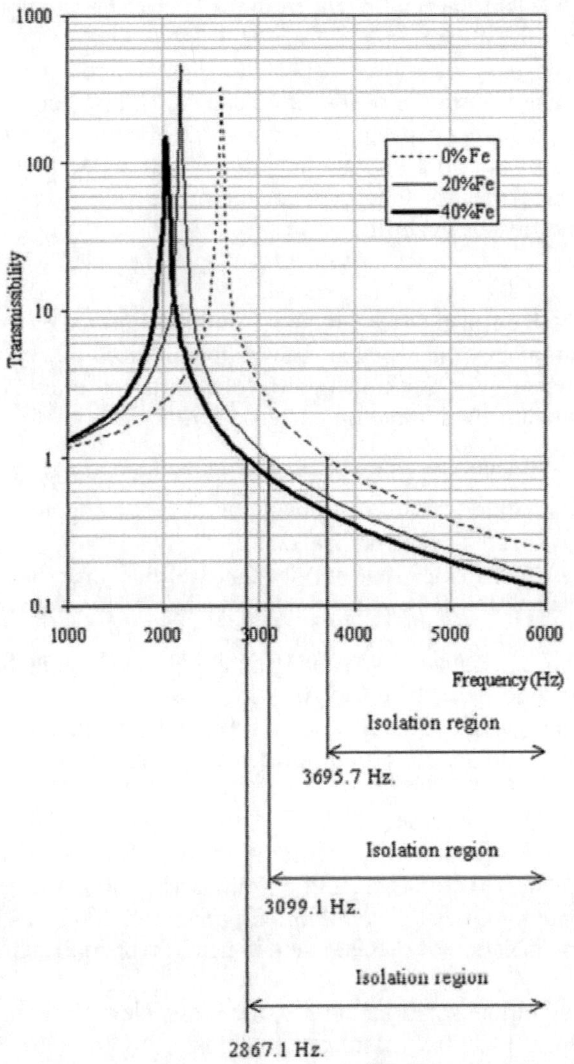

Figure.4. The graph of transmissibility of the product and range of isolation region.

It can be obtained from equation 4, the minimal operating frequency for isolation region for each specimen. As can be seen in figure. 4, for specimen with no addition of gray cast iron the value of minimum operating frequency is 3695.7 Hz. Addition about 20 vol.% gray cast iron make the minimal operating frequency decrease to become 3066.1 Hz. The isolation region was very much increase with addition of 40 vol.% gray cast iron that was caused by the decrease of minimal operating frequency to become 2867.1 Hz. These all mean that addition of gray cast iron can extend the isolation region.
Comparing with transmissibility of gray cast iron from the previous research [8], it was found that the product with addition 40 vol.% gray cast iron seem have similar transmissibility with gray cast iron.
It should be note that the metal composite in this research is dedicated for passive vibration isolator with limitation that the system will work if frequency of the disturbance is around the natural frequency of the composite[9].
In this research, waste of grey cast-iron is used to make a composite wit aluminium. Another waste for future investigation is ductile cast iron. Ductile cast iron will have good properties of damping capacity trough process of austempered[10]. The Damping capacity data from various metal and

ceramic is well established[11]but the final result from the composite can not be predicted and should be obtained by experimental investigation.
Beside grey cast iron, shape memory alloys, especially those based on copper, present a different damping capacity on martensite, austenite or transition state [12]. It is good for future investigation to develop new type of passive damping capacity by using waste of copper chips.
This is one of the example success work in manufacturing useful product by utilization waste of metal following previous works [13,14]. It is suggested for the future work to promote recycle, reuses and reduce waste of metal for future sustainability.

4. Conclusion
The conclusions that is obtained in this researchare :Addition gray cast iron affect on reducing the natural frequency. The damping capacity decrease with addition 20 vol.% gray cast iron, and conversely increase significantly by addition of 40 vol.% gray cast iron.The isolation region was extended and transmissibility at isolation region decreased with addition of gray cast iron.

5. References
[1] Baik, S. H.,High damping Fe-Mn martensitic alloys for engineering application, *Nuclear Engineering and Design*, 2000, 198, pp. 241-252.
[2] Lee, D.O and Han, J.H., A Comparison of Vibration Isolation Characteristics of Various Forms of Passive Vibration Isolator, *The Korean Society for Noise and Vibration Engineering*, 2012, Vol. 22, Issue 9, 2012, pp.817-824.
[3] Shenglong, D., Dabo,L., Tianzhen,W. and Chunyu, L., Damping behavior and mechanical properties of rapidly solidified Al-Fe-Mo/Al Alloys, *Journal of Material Science*, 1998 33, pp. 2227-2231.Li,P.Y., Dai,S. L., Chai,S. C., and Li,Y. R, High damping Al-Fe-Mo-Si/Zn-Al composite produced by rapidly solidified powder metallurgy process, *Scripta Mater*,2000, 42, pp. 955-960.
[4] Li,P.Y., Dai,S. L., Chai,S. C., and Li,Y. R, High damping Al-Fe-Mo-Si/Zn-Al composite produced by rapidly solidified powder metallurgy process, *Scripta Mater*,2000, 42, pp. 955-960.
[5] Xie,Z., Tane, M., Hyun, S., Okuda, Y., and Nakajima H., 2006, Vibration-damping capacity of lotus porous magnesium, *Material Science and Engineering*, 2006, A 417, pp. 129-133.
[6] Rivin, E. I., Vibration isolation of precision equipment, Precision Engineering, 1995, 17, pp. 41-56.
[7] Rivin, E.I., Passive vibration isolation, 2003 Asme Press, New York, p. 35.
[8] Walton, C. F.and Opar, T. J., Iron castings handbook, 1981 Iron casting society, Inc., USA, p. 167.
[9] Patil, M.S., Hada, M.K., Bhave, S.Y., and Joshi, S.G., Vibration Isolation and TransmissibilityCharacteristics of Passive Sequential Damper Defence Science Joumal, Vol. 54, No. 1, January **2004, pp.** 39-51.
[10] Kang, C.Y., Sung, J.H., Kim, G.H., Kim, B.S. and Kim,I.S., Effect of Heat Treatment on the Damping Capacityof Austempered Ductile Cast Iron, *Materials Transactions*,2009, Vol. 50, No. 6, pp. 1390- 1395.
[11] Zhang, J., Perez, R.J., Llavernia, E.J., Documentation of damping capacity of metallic, ceramic and metal-matrix composite materials, Journal of Materials Science, 1993, 28, pp.2395-2404.
[12] Crăciun, R.C, Stanciu, S., Cimpoeşu,R., Ursanu, A.I., Manole, V., Paraschiv, P. and Chicet, D.L., Metallic materials for mechanical damping capacity Applications, *Materials Science and Engineering*, 2016,147, pp.1-8.
[13] Nindhia, T.G.T, Surata, I W., Atmika, I K. A, Negara, D.N.K.P., and Putra, G.P.A.L., Biogas Desulfurizer Made from Waste of Aluminium Chips, *International Journal of Materials, Mechanics and Manufacturing*, Vol. 2, No. 3, August 2014, pp. 219-222.
[14] Nindhia, T.G.T, Sucipta, I M., Surata, I W., Adiatmika, I K.A., Negara, D.N.K.P, Negara, K.M.T., Processing of Steel Chips Waste for Regenerative Type of Biogas Desulfurizer, *International Journal of Renewable Energy Research*, Vol.3, No.1, 2013, pp. 84-87

IOP Conf. Series: Materials Science and Engineering **540** (2019) 012010 doi:10.1088/1757-899X/540/1/012010

Synthesis of Montmorillonite Nanoclay Reinforced Chitin-cellulose Nanocomposite Film

Tiffany Louise Lao, Lance Tristan Pengson, Jordan Placido and Leslie Joy Diaz

Department of Mining, Metallurgical and Materials Engineering, University of the Philippines, Diliman, Quezon City, Philippines
E-mail: tblao@up.edu.ph

Abstract. Montmorillonite (MMT) nanoclay reinforced chitin-cellulose nanocomposite is a biodegradable polymer that has potential as a food packaging material. In this study, it was synthesized using 1 and 5 wt% MMT (1NC and 5NC) via solution casting. Good miscibility of chitin and cellulose was interpreted by the presence of cylindrical particles revealed in the SEM image, and the XRD peak at ~19° and shoulder at ~23°. Peaks were also observed at 2.62 and 2.27° for 1NC and 5NC, respectively, that corresponded to 16.87 and 19.47 Å interlayer spacings, both of which were higher than that of pristine MMT which is 12.57 Å. This suggested that chitin-cellulose penetrated MMT and that 5NC had a lower degree of intercalates, which was confirmed by EDX mapping. This better MMT dispersion was speculated to be due to the nematic state of 5NC gel. With molecules sliding off easily in the gel, polymer penetration in the interlayer spacing was easier. This was verified by its lower apparent viscosity. Finally, this better MMT dispersion resulted to a good stress transfer and, maybe, introduced a more torturous crack path for 5NC. These improved its toughness (2.22 J/mm3) and Young's modulus (6. 25 GPa).

1. Introduction

The world has generated of at least 6.3 billion metric tons of plastic waste in which ~79% of it is accumulated in landfills or natural environment. Without action, an increase to 12 billion metric tons of landfill plastic waste is expected by 2050 [1]. This huge amount of plastic waste is highly-contributed by single-used plastic, or throwaway. This includes food packages.

To address this, various environmental sustainability practices have already been implemented or promoted by the government. One of these is the usage of paper-based containers as a substitute to petrochemical-based polymeric counterparts. However, this seemingly good alternative is, in fact, further threatens the sustainability of the environment. This is because of: (1) virgin paper pulp mass production that promotes tree felling; and (2) the polyethylene film, which is used for inside proofing, is non-recyclable and takes years to degrade. For these reasons, fabrication of food packaging material that utilizes biodegradable polymer is a topic of research nowadays.

Chitin-cellulose film is a biodegradable polymer that has potential to be used as a food packaging material [2]–[3]. However, its water vapor permeability (WVP) is still high to be used for such application. For this reason, montmorillonite (MMT) nanoclay was added to it, which resulted to a decrease of WVP [4]. But, the film became brittle in the process due to the incorporation of immense amount of MMT. Therefore, for this study, MMT nanoclay was added in smaller amounts to enhance the mechanical integrity of this film.

2. Methodology

2.1. Materials
Chitin flakes (C9213), cellulose medium fibers (C6288), montmorillonite nanoclay (Nanomer 1.34MN, 682640) and N,N-dimethylacetamide (185884) were obtained from Sigma-Aldrich. Lithium chloride (Dried, Unilab 292-500) was procured from Ajax Fine Chem. Methanol and isopropanol were acquired from Pharmco-Aaper and Macron, respectively.

2.2. Fabrication of nanocomposite film
Chitin was dissolved in lithium chloride-dimethylacetamide solution (LiCl-DMAc). MMT was then added to it and sonicated. The resulting slurry was stirred to improve the homogenization. Two different concentrations of MMT, 1 wt% and 5 wt%, were used. Cellulose was swelled using deionized water (DI), methanol and DMAc. After, it was also dissolved in LiCl- DMAc.
The MMT-chitin mixture and cellulose solution was combined and sonicated, and then casted into a glass tray forming a gel. The gel was soaked in isopropanol then methanol. Excess solvent was removed. After soaking, it was cold-pressed and then oven dried. This resulted to a nanocomposite film with average thickness of 0.022 mm.

2.3. Determination of microstructure of nanocomposite film
Microstructure of the nanocomposite films were observed via a field emission scanning electron microscope (FESEM, Hitachi SU-8230) with an accelerating voltage of 4 kV. Magnifications of 500, 10kx and 20 kx were used. Samples were sputter coated with Pt for 60 sec prior to the test. Image J was utilized to measure the dimension of the chitin-cellulose particles.
MMT dispersion was verified using energy dispersive X-ray spectrometer (EDX) with an accelerating voltage of 5 kV to measure the Si-Kα signal. X-ray diffractometer (XRD, Rigaku SmartLab) was also employed using a Cu-Kα target with a scan rate of 2°/min.

2.4. Mechanical integrity of film
Following ASTM D882-12, tensile test was conducted using a Universal Testing Machine (UTM, Shimadzu Autograph-IC) with 1 kN load cell to measure Young's modulus. Origin Pro8 was utilized to determine the toughness of the films via area integration of the stress-strain curves obtained from the tensile test.

2.5. Supplementary tests
Using the same parameters, SEM and XRD were conducted to verify the miscibility of chitin and cellulose. Moreover, using Brookfield DV-II + Pro with spindle no. 63, viscosities of the MMT-chitin-cellulose mixtures were measured to validate the MMT dispersion result.

3. Results and discussion
The chitin-cellulose films with 1 wt% and 5wt% montmorillonite (MMT) nanoclay (1NC and 5NC, respectively) were transparent as shown from Figure 1. This was attributed to the amount of MMT incorporated into the film. In contrast, [4] produced a brown-colored and opaque film, because of the large incorporation of MMT, that was 180 wt%.
It was also observed that folds were present at the sides for both films, as indicated by the arrows in Figure 1. These folds were caused by the shrinkage of the films during the removal of lithium chloride-dimethylacetamide (LiCl-DMAc) in the soaking phase [3]. The hydroxyl proton of cellulose (and also maybe chitin) is bonded with the Cl^- in which is accompanied by the Li^+ that is solvated by the DMAc molecules [5]. When the gel was soaked with isopropanol and methanol, DMAc was dissolved to these alcohols. This would have disrupted the bonding of chloride of LiCl and hydroxyl proton of cellulose (and also maybe chitin), leaving only the hydrogen-bonded chitin-cellulose [6].
Finally, few small dots that were non-uniformly dispersed were also noticed in both films, see red circles in Figure 1. These could be the remnants of few undissolved chitin and/or cellulose. Nanoclay being the small visible dots was not plausible, because of its very low concentration and uniform

dispersion in the chitin-cellulose matrix, as represented by the uniformly-dispersed Si atoms shown in Figure 3c and d.

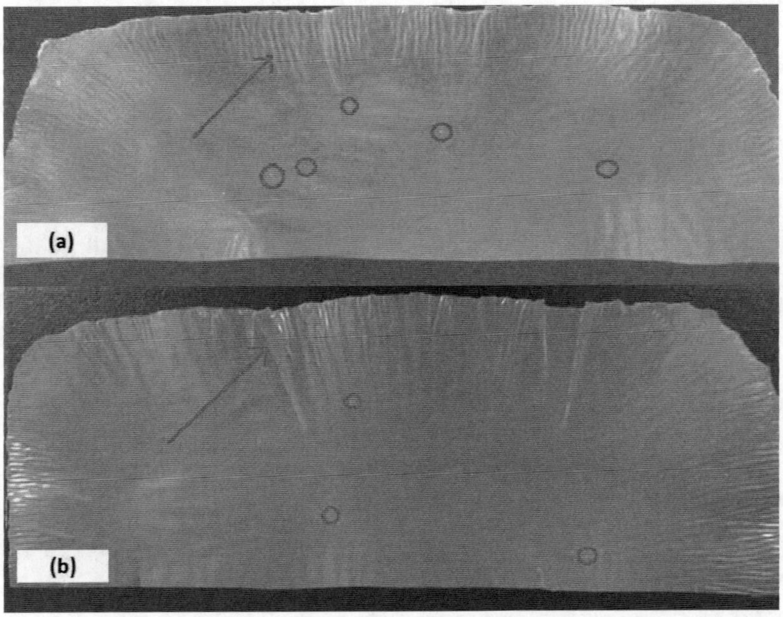

Figure 1. Visual observation of (a) 1 wt% and (b) 5 wt% montmorillonite nanoclay reinforced chitin-cellulose nanocomposite films.

3.1. Microstructure of the nanocomposite film

As shown in the micrographs of Figure 2a and b, mounds and fibrous-like structures were present at the surface of both 1 and 5NC films. The latter could be attributed to cellulose because of its fiber form [3]. However, the former could be inferred as chitin. These microheterogeneities of chitin-cellulose blend is expected, because of the negative Flory-Huggins parameters which were calculated at cellulose content greater than 2.5wt%, and, in this study, 20% was used [7]. Nevertheless, accumulation of these structures at certain sites of the films could have resulted to the appearance of small dots in Figure 1.

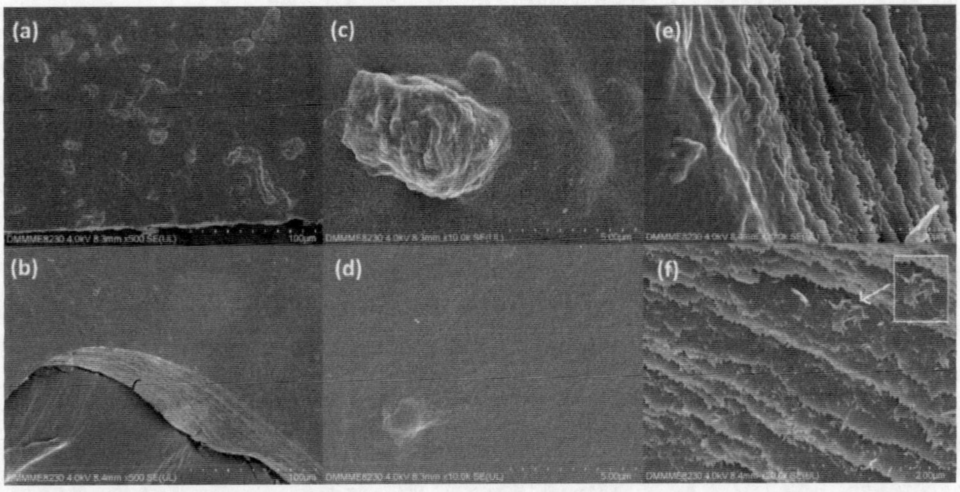

Figure 2. Comparison of surface, surface mounds, and cross-section of 1 wt% (a, c, e) and 5 wt% (b, d, f) montmorillonite nanoclay reinforced chitin-cellulose nanocomposite films.

IOP Conf. Series: Materials Science and Engineering **540** (2019) 012010 doi:10.1088/1757-899X/540/1/012010

It should be noted that a comparison of the degree of surface roughness induced by the said microheterogeneities for 1 and 5NC would be difficult to imply for this study. The seemingly "smooth" surface shown for 5NC in the Figure 2d micrograph is only applicable for certain areas. It is because 5NC had also large mounds with size comparable to that of 1NC (not shown). Furthermore, the surface of both films as shown by the micrographs generated by SEM-EDX, see Figure 3a and b, showed equivalent qualitative roughness.

For the cross-sections of the films, layers were observed for both films. This was also the case for the study of Zhang for chitin-cellulose in 2002 [8]. However, it should not be inferred that this is an inherent structure of chitin-cellulose. It is because the structure could be the effect of cold-pressing. Moreover, if focused in the Figure 2f inset, cylindrical particles of diameter of 37.2 nm appeared as the components of each layer. These were the chitin-cellulose particles [6]. This was validated by the peak at ~19° with a shoulder at ~23° present in the diffractograms of both 1 and 5NC films shown in Figure 5. This indicated a good miscibility of cellulose and chitin [9].

In terms of nanoclay distribution in the chitin-cellulose film, SEM-EDX mapping was conducted. Silicon atoms of MMT was used as the determining atoms to describe distribution, because: (1) oxygen atoms of MMT would be impossible to pinpoint due to the fact that both cellulose and chitin have also oxygen, and (2) the number of aluminum atoms is only a half of that of silicon, which, if chosen as the determining atom, would be more difficult to magnify the nanoclay presence, due to the low nanoclay content of the films.

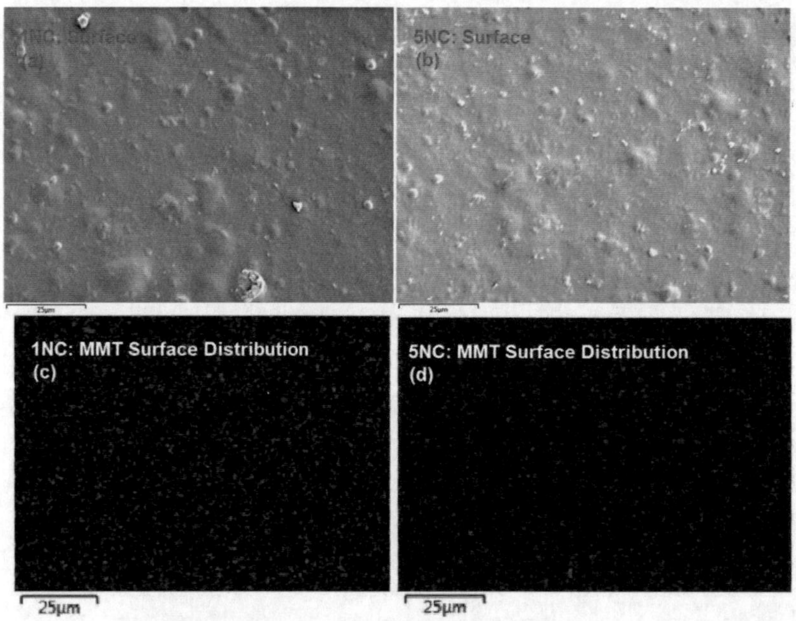

Figure 3. SEM-EDX mapping of monmorillonite nanoclay surface distribution in 1 wt% and 5 wt% montmorillonite nanoclay reinforced chitin-cellulose nanocomposite films.

Seen in Figure 3c and d, both films had areas of agglomeration, as shown by the red patches, which implied that intercalates, or, maybe, tactoids, were the abundant nanoclay dispersed structures [10]. Comparing the two films, 1NC was more agglomerated than 5NC. Agglomeration could have resulted, because the sonication and mechanical stirring were not enough to break the clay-clay surface interactions [11]. This was verified by the left shift from 2.62 to 2.27° peak for the MMT signal of 5NC, see Figure 5, which, using Bragg's equation, would be the interlayer spacing increase from 16.87 to 19.47 Å. Both were higher than the interlayer spacing of pristine MMT that was 12.57 Å, which would mean a successful penetration of chitin-cellulose into the MMT. Additionally, this polymer penetration could have decreased the intra-chain bonding of chitin, which corresponded to the lowered intensity of its (013) plane for 5NC.

Better dispersion of MMT in the chitin-cellulose matrix in 5NC may be due to the nematic state degree of its gel. In this study, pristine chitin/LiCl-DMAc solution was already in this state, because it exhibited a glittering light yellow visual [12]. With this, it was speculated that addition of 5 wt% MMT further favored the solution to become more nematic. Since molecules slide off with each other the easiest in this state [12], polymer penetration to the interlayer spacing was easier for 5NC. This was verified by its lower apparent viscosity value as shown in Figure 4.

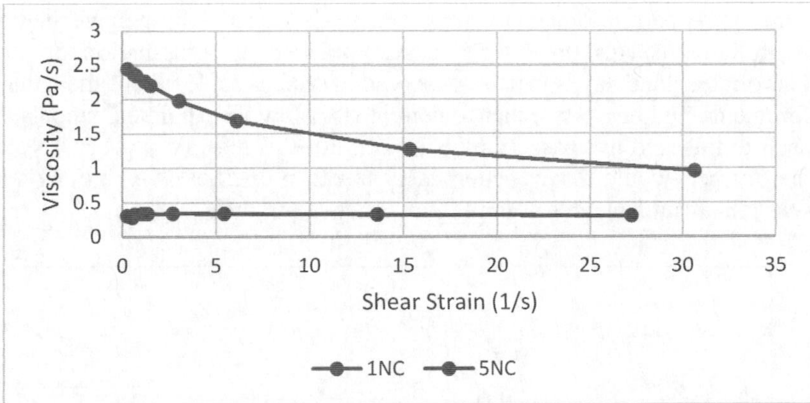

Figure 4. Viscosity measurements of 1 wt% and 5 wt% montmorillonite nanoclay reinforced chitin-cellulose nanocomposite films in increasing shear rate.

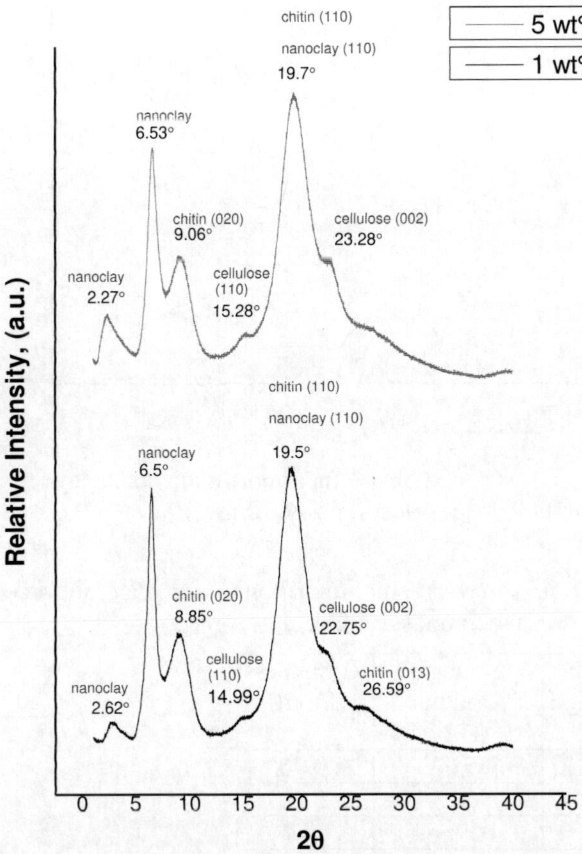

Figure 5. X-ray diffractograms of 1 wt% and 5 wt% montmorillonite nanoclay reinforced chitin-cellulose nanocomposite films.

IOP Conf. Series: Materials Science and Engineering **540** (2019) 012010 doi:10.1088/1757-899X/540/1/012010

3.2. Mechanical strength of nanocomposite film

Nanoclay with an exfoliated dispersion structure in a polymer matrix would result to a composite with higher mechanical properties than with an intercalated counterpart [13], [14]. This enhancement is caused by: (1) the high aspect ratio that enables high nanoclay surface area to interact with the polymer matrix, and (2) the better nanoclay dispersion and distribution in the said matrix, which results to a uniform matrix-to-reinforcement stress and strain transfer throughout the composite [14]. Since complete exfoliation seems impossible in real practice, a combination of phase-separated, intercalated, and exfoliated nanoclay is usually observed in which the degree of the latter is maximized to achieve the aforementioned property enhancement [15]. A low incorporated nanoclay concentration usually would result to this said maximization [16]–[18]. For this reason, between 1NC and 5NC, it is expected that the former would have better mechanical properties, i.e. Young's modulus and toughness. However, the results in table 1 and Figure 6 begged to differ.

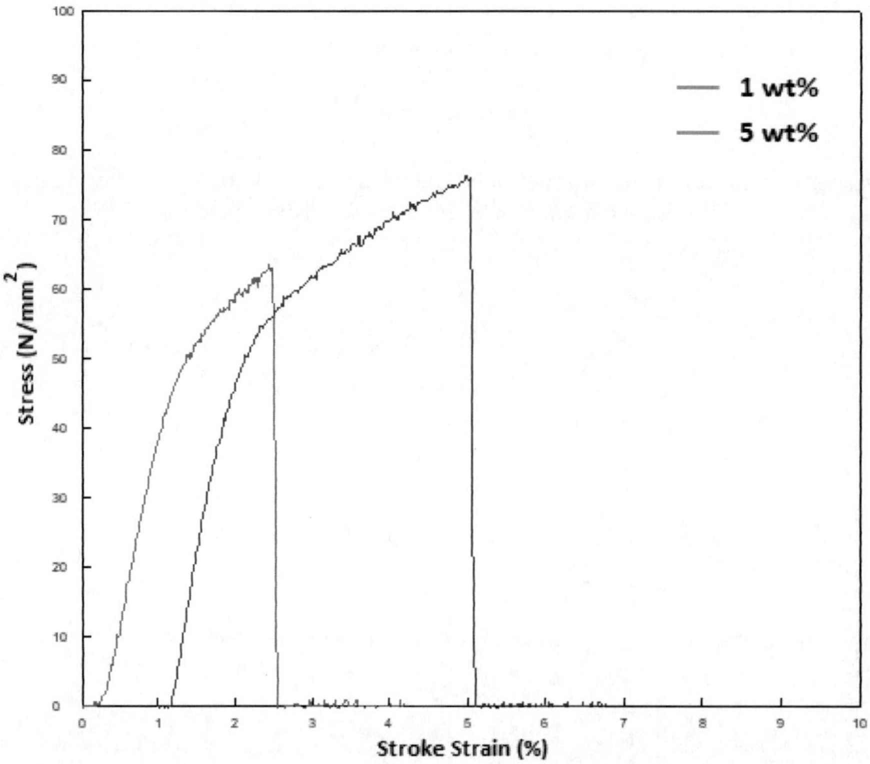

Figure 6. Stress-strain curve of 1 wt% and 5 wt% montmorillonite nanoclay reinforced chitin-cellulose nanocomposite films.

Table 1. Young's moduli of 1 wt%, 5 wt%, and 1 (w/v)% montmorillonite nanoclay reinforced chitin-cellulose nanocomposite films.

Nanoclay Concentration	Young's Modulus (GPa)	Toughness (J/mm^3)
1 wt%	5.67	0.96
5 wt%	6.25	2.22
1 (w/v)%	0.03[a]	--
[a] [3]		

Increase of clay concentration should have incorporated brittleness, because it is a ceramic. However, it was determined through integration of the stress-strain curves in Figure 6 that 5NC was 43.24% tougher than 1NC. It should be noted that, because of the increase in interlayer spacing shown in Figure 5, nanoclay was better distributed in 5NC as verified by the lower amount of agglomerates shown in the EDX maps in Figure 3. This better nanoclay distribution might have introduced a torturous path for crack propagation which resulted to a tougher 5NC [13], [16]. On the other hand, 1NC was less tough, because its larger amount of agglomerates could have acted as stress concentrators resulting to a premature film breakage, instead of crack inhibitors [13], [17].
Meanwhile, for the Young's moduli of 1NC and 5NC, it was determined that the latter had the larger value. Evidently, it should have the higher value, because it had a wider interlayer spacing. And, as mentioned, wider interlayer spacing would result to a well dispersed and distributed nanoclay in the film that would result to a better stress (or strain) transfer between the chitin-cellulose matrix and the nanoclay reinforcement, consequently, caused a higher Young's modulus. However, in contrast to the toughness values difference, the difference of Young's moduli between the films was only 8.97%. This could have resulted from the small interlayer spacing difference of these films [12]. Furthermore, using the data of [3], it is observed that increasing the nanoclay concentration further to 1 (w/v)% resulted to a drop of the said modulus [3], [12]. This might have caused by the 30.97% decrease of the interlayer spacing at the said nanoclay content, which could have signified higher degree of agglomeration. By integrating this Young's Modulus-interlayer spacing relationship, it could be inferred that, maybe, with a better dispersion-distribution technique, the nanoclay content that could result to the highest degree of exfoliation, if not 5wt%, would be near to that concentration.

4. Conclusion

Montmorillonite (MMT) nanoclay reinforced chitin-cellulose nanocomposite film (NC) was successfully fabricated using 1 and 5wt% of MMT. It was determined that the latter was tougher (2.22 vs. 0.96 J/mm3) and stiffer (Young's modulus: 6.25 vs. 5.67 GPa) than that of 1 wt%. This was due to the lower nanoclay agglomeration experienced by the said film, as shown via EDX mapping, which was further proven by the increase of nanoclay interlayer spacing from 16.87 to 19.47 Å. It is suggested that low agglomeration of the latter film was caused by the nematic state of its gel, which made the molecules in it to slide off easily with each other. This was verified by its lower viscosity.
Moreover, good miscibility of chitin and cellulose solution was proven through the 19.7° chitin peak with a 23.28° cellulose shoulder shown in the X-ray diffractograms. Thus, even if surface mounds were present in both films, it could be stated that microheterogeneity of chitin and cellulose was only at a lower extent as to not mask the said peak associated with the said miscibility.

References

[1] Geyer R, Jambeck J R and Law K L 2017 Production, use, and fate of all plastics ever made. *Sci. Adv* **3**

[2] Poblete M R and Diaz L J L 2014 Synthesis of biodegradable cellulose-chitin polymer blend from portunus pelagicus *Adv. Mater. Res.* **925** 379–84

[3] Manuzon M J, Macailing M and Diaz L J L 2018 Lamination of chitin-cellulose nanocomposite and thermoplastic starch *Key Eng. Mater* **775** 81–8

[4] Macailing M and Diaz L J L 2015 *Synthesis and characterization of chitin-cellulose nanocomposite laminates* (Quezon City, Philippines: University of the Philippines graduate dissertation)

[5] Zhang C, Liu R, Xiang J, Kang H, Liu Z and Huang Y 2014 Dissolution mechanism of cellulose in N,N-dimethylacetamide/lithium chloride: revisiting through molecular interactions *J. Phys. Chem.* B **118** 9507–14

[6] Wu J, Liang S, Dai H, Zhang X, Yu X, Cai Y, Zhang L, Wen N, Jiang B and Xu J 2010 Structure and properties of cellulose/chitin blended hydrogel membranes fabricated via a solution pre-gelation technique *Carbohydr. Polym.* **79** 677–84

[7] Nud'ga L A, Petrova V A, Bochek A M, Pretropavlovskii G A, Alekseev V L, Evmenenko G A and Kalyuzhnaya L M 1999 Structure of chitin-cellulose mixtures in solution and in the state *Vysokomolekularnye Soedineniya. Ser.A Ser.B Ser.C - Kratkie Soobshcheniya* **41** 1791–2

[8] Zhang L, Guo J and Du Y 2002 Morphology and properties of cellulose/chitin blends membranes from NaOH/thiourea aqueous solution *J. Appl. Polym. Sci.* **86** 2025–32

[9] Zhou D, Zhang L and Guo S 2005 Mechanisms of lead biosorption on cellulose/chitin beads *Water Res.* **39** 3755–62

[10] Selvakumara V, Palanikumar K and Palenivelu K 2010 Studies on mechanical characterization of polypropylene/Na+-MMT nanocomposites *JMMCE* **9** 671–81

[11] Bensadoun F, Kchit N, Billote C, Trochu F and Ruiz E 2011 A comparative study of dispersion techniques for nanocomposite made with nanoclays and an unsaturated polyester resin *J. Nanomater.* **2011**

[12] Chen B, Sun K and Zhang K 2004 Rheological properties of chitin/lithium chloride, N,N-dimethylacetamide solutions *Carbohydr Polym.* **58** 65-9.

[13] Majdzadeh-Ardakani K, Navarchian A H and Sadeghi F 2010 Optimization of mechanical properties of thermoplastic starch/clay nanocomposites *Carbohydr. Polym.* **79** 547–54.

[14] Stiller B 2008 *The effect of montmorillonite nanoclay on mechanical and barrier properties of mung bean starch films* (Clemson, SC: Clemson University graduate dissertation)

[15] Nguyen Q T and Baird D G 2007 An improved technique for exfoliating and dispersing nanoclay particles into polymer matrices using supercritical carbon dioxide *Polym. J.* **48** 6923–33

[16] Castro-Aguirre E, Auras R, Selke S, Rubino M and Marsh T 2018 Impact of nanoclays on the biodegradation of poly(lactic acid) nanocomposites Polym. J. **10**

[17] Bashar M, Mertiny P and Sundararaj U 2014 Effect of nanocomposite structures on fracture behavior of epoxy-clay nanocomposites prepared by different dispersion methods *J. Nanomater.* 2014

[18] Lai C Y, Groth A, Gray S and Duke M 2014 Preparation and characterization of poly(vinylidene fluoride)/nanoclay nanocomposite flat sheet membranes for abrasion resistance *Water Res.* **57** 56–66

Acknowledgments

Authors would like to thank the Department of Mining, Metallurgical, and Materials Engineering of the University of the Philippines for sponsoring the raw materials and chemical reagents used in this study.

IOP Conf. Series: Materials Science and Engineering **540** (2019) 012011 doi:10.1088/1757-899X/540/1/012011

Experimental Characterisation and Numerical Modelling of the Effect of Cold Rolling on the Nanoindentation Response of Pure Zinc Grains

P T N Nguyen[1,2], F Abbès[2], B Abbès[2], J-S Lecomte[3] and C Schuman[3]

1 The University of Danang, University of Science and Technology, Da Nang, Vietnam
2 GRESPI, University of Reims Champagne-Ardenne, France
3 LEM3, University of Lorraine, Metz, France
E-mail: fazilay.abbes@univ-reims.fr

Abstract. In this study, the orientation-dependent response of as-received annealed cold-rolled pure zinc and material with thickness reduction rate of 50% grains using instrumented indentation tests is investigated. The experiments were characterized by orientation microscopy and atomic force microscopy scans to quantify the orientation-dependent mechanical response during nanoindentation. The single crystal hardening parameters are then identified for each family of slip system by using crystal plasticity finite element (CPFE) simulations. Comparison between experimental and numerical results in terms of "load-penetration depth" curves show a good agreement. The increased percentage of cold reduction increases the identified critical resolved shear stress (CRSS). Finally, the accuracy of the model is evaluated by comparing experimental and numerical data issued from nanoindentation response grains of distinct crystalline orientations involving different slip systems activity rates.

1. Introduction

Zinc is well-suited for corrosion protective coating for iron and steel products. It is successfully used as a protective coating on a variety of products and in many exposure conditions, especially in automotive applications (hot-dip galvanized sheets) for its high corrosion resistance in most environments. An efficient coating requires a good forming ability and adhesive bonding, with no crack or flaking during forming process. Zinc has a hexagonal close-packed (HCP) crystalline structure. Zinc is characterized by a high lattice ratio exhibiting a strong anisotropic behaviour compared to other HCP metals like titanium or magnesium $\left(\frac{c}{a}\Big|_{Zn}=1.856,\ \frac{c}{a}\Big|_{Mg}=1.624,\ \frac{c}{a}\Big|_{Ti}=1.588\right)$.

HCP sheet forming finite element simulations have been generally performed using classical macroscopic anisotropic criteria [1, 2]. A physically based approach requires a good understanding of the deformation mechanisms: dislocation slip at the individual grain scale in the polycrystalline sample. Moreover, the grain response involves different slip systems activity rates depending on its orientation. Recent advances in the developing of macroscopic plasticity models for HCP materials have allowed a better capture of specific features like the anisotropy and the tension-compression asymmetry in yielding of HCP metals [3-8]. An accurate determination of slip resistance and its evolution is of very importance to well predicting the macroscopic response of a polycrystalline material. In practice, instrumented indentation tests performed on grains of distinct crystallographic orientations allows determining indirectly the critical resolved shear stresses (CRSS) of all systems.

IOP Conf. Series: Materials Science and Engineering **540** (2019) 012011 doi:10.1088/1757-899X/540/1/012011

To achieve this, an inverse parameter identification procedure combining crystal plasticity finite element method (CPFEM) with an optimization algorithm is needed [9-11]. In this study, material parameters were determined by inverse method: finite element simulations of the nanoindentation tests are used, and the parameters are varied systematically until agreement between simulations and experiments is achieved. The objective function is the deviation between experimental and numerical results in terms of "load-penetration depth" curves issued from the nanoindentation experiments performed on grains of given crystalline orientation. Cold-rolled grains of pure polycrystalline zinc sheets are considered. The identified CPFEM parameters for materials with different reduction rates are compared.

2. Crystal Plasticity model

In the crystal plasticity theory, the plastic deformation results from the cumulative effect of crystalline glide in all activated slip systems. The deformation rate is the sum over all the shear rates contributed by all activated systems. The shear strain rate $\dot{\gamma}^{(\alpha)}$ on the αth slip system obeys Schmid's law and is given by the following power-law:

$$\dot{\gamma}^{(\alpha)} = \dot{\gamma}_0{}^{(\alpha)} \left| \frac{\tau^{(\alpha)}}{\tau_c{}^{(\alpha)}} \right|^n sgn\left(\tau^{(\alpha)}\right) \tag{1}$$

where $\dot{\gamma}_0{}^{(\alpha)}$ is the reference shear strain rate on slip system α, $\tau^{(\alpha)}$ is the resolved shear stress, $\tau_c{}^{(\alpha)}$ is the slip resistance or critical resolved shear stress of the slip system α, and n is the strain-rate sensitivity exponent.

For multiple slip, the evolution of the slip resistance $g^{(\alpha)}$ is governed by the hardening law expressed as:

$$\dot{g}^{(\alpha)} = \sum_{\beta} h_{\alpha\beta} \dot{\gamma}^{\beta} \tag{2}$$

where $h_{\alpha\beta}$ are the slip hardening moduli, the sum ranges over all activated slip systems. $h_{\alpha\alpha}$ and $h_{\alpha\beta}$ ($\alpha \neq \beta$) are the self and latent hardening moduli, respectively.

Peirce *et al.* [12] have expressed the relation between self and latent hardening moduli as follows:

$$\begin{cases} h_{\alpha\alpha} = h(\gamma) = h_0 \sec h^2 \left| \dfrac{h_0 \gamma}{\tau_s - \tau_0} \right| \\ h_{\alpha\beta} = q h(\gamma) \quad (\alpha \neq \beta) \end{cases} \tag{3}$$

where h_0 is the initial hardening modulus, τ_0 is the critical resolved shear stress and τ_s is the saturation strength. γ is the Taylor cumulative shear strain on all slip systems defined as:

$$\gamma = \sum_{\alpha} \int_0^t \left| \dot{\gamma}^{(\alpha)} \right| dt \tag{4}$$

The parameter q is defined as the ratio of latent over self-hardening moduli and characterizes the hardening behaviour. It, taken equal to unity in our simulations.

The basal glide ($B\langle a \rangle$ $\{0001\}\langle 11\bar{2}0 \rangle$) is the primary slip system for zinc because of its high lattice ratio. The overall deformation is accommodated by prismatic glide ($P\langle a \rangle$ $\{10\bar{1}0\}\langle 11\bar{2}0 \rangle$) and second-order pyramidal glide ($\Pi_2\langle c+a \rangle$ $\{11\bar{2}2\}\langle 11\bar{2}3 \rangle$) to satisfy the Taylor criterion [13] requiring five independent slip systems for the ductility of polycrystalline metals. It is worth mentioning that we have not considered first-order pyramidal slip systems ($\Pi_1\langle a \rangle$). Indeed, the analysis of Yoo and Wei [14] showed that basal and prismatic slip systems are cristallographically equivalent to $\Pi_1\langle a \rangle$ glide. Following the work of Yoo and Lee [15], zinc exhibits compression twinning $\{10\bar{1}2\}\langle 10\bar{1}\bar{1} \rangle$ causing a

contraction along the c-axis. However, no twinning was observed in the nanoindentation experiments we have carried out, therefore the compression twin system is not considered in this study.
The crystal plasticity model for an HCP crystal was implemented into the implicit finite element (FE) code ABAQUS/Standard through a user material subroutine (UMAT), initially developed by Huang [16] for cubic system and modified by Kysar and Hall [17]. In the FE formulation, the plastic deformation is assumed due solely to the crystallographic dislocation glide. The deformation by diffusion, twinning and grain boundary sliding is not considered here.

3. Instrumented indentation tests

The first studied material is the as-received annealed cold-rolled sheet (1 mm thickness) of high-purity polycrystalline zinc (Zn > 99.99 + wt%) named "AR". The chemical composition is given in table 1. The second studied material is the as-received sheets that have been plastically deformed (thickness reduction rate of 50%) named "TR" and annealed at a temperature of 100°C for a period of 2 h to trigger the recrystallization of new grains.

Table 1. Chemical analysis (ppm) of as-cast high purity (99.99+ %) polycrystalline zinc.

Ag	Cd	Ca	Cr	Cu	Cs	Fe	Mn	Ni	Pd	Pb	Rh	Ta	T	Ti
0.36	3.4	<0.1	0.55	4.0	<0.1	4.9	0.18	0.66	<1	12	<0.5	<10	5.1	4.5

Microstructure characterization was evaluated by electron backscatter diffraction (EBSD) using a JEOL65 field emission gun-scanning (FEG) electron microscope (SEM) equipped with OIM EBSD system. Samples preparations for EBSD examination were first ground with sand paper, and then were polished with 1μm diamond paste to mirror finish. In order to achieve the surface quality required for nanoindentation tests and EBSD examination (*i.e.* deformation-free and scratch-free surface) a chemo-mechanical polishing with colloidal silica suspension (~ 40nm) on porous neoprene polishing cloth for about 35nm was used.
Instrumented indentation tests were performed using a 90° cono-spherical diamond indenter (tip radius of 1 μm) mounted in a NHT-2 nanoindenter (CSM Instruments). Indentations were carried out in load-control mode.
The selected grains for nanoindentation tests are located in the standard stereographic triangle as shown in figure 1. The crystalline orientation using the Euler-Bunge angles of those grains are listed in table 2. The grains for which the <c> axis (*i.e.* <0001>) is parallel to the normal direction of the sheet are called basal grains. Grains for which the $\langle \bar{1}2\bar{1}0 \rangle$ direction is parallel to the normal direction of the sheet are called pyramidal grains. Finally, grains for which the $\langle 01\bar{1}0 \rangle$ direction is along the normal direction of the sheet are called prismatic grains. To avoid grain boundary influence, indents were located in the mid-zone of individual grains.

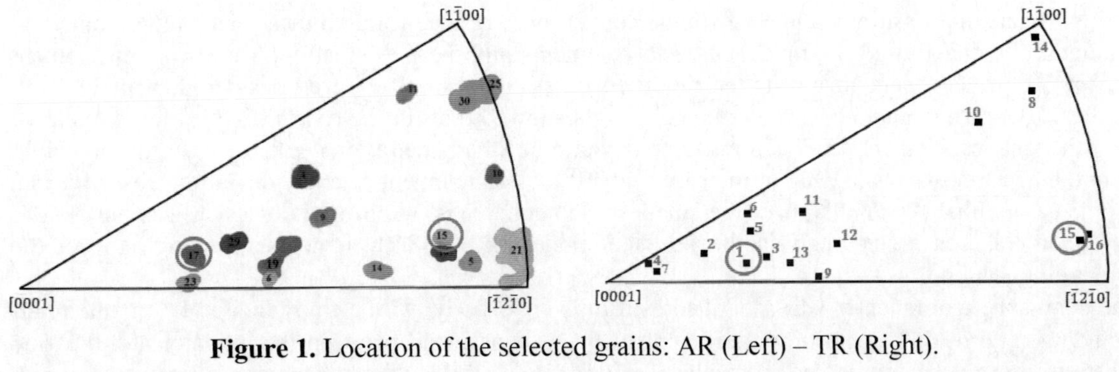

Figure 1. Location of the selected grains: AR (Left) – TR (Right).

Table 2. Crystalline orientation of the selected grains.

Grain number	Crystalline orientation	State	Euler-Bunge angle (°)		
			φ_1	ϕ	φ_2
1	Basal	TR	2.4	23.3	37.6
15	Pyramidal	TR	50.3	83.8	24.7
17	Basal	AR	114.9	28.4	189.8
15	Pyramidal	AR	69.6	107.1	186.3

4. Identification process

Three-dimensional finite element simulations of the nanoindentation tests were performed using the commercial software ABAQUS. Assuming a single-crystal-like behaviour, the grain has been modelled as a cylinder of 20 μm diameter and 20 μm height. The sample was discretized using 13860 eight-node linear hexahedral elements with reduced integration scheme (C3D8R), and the mesh size with the smallest element size equals to 60 nm to ensure that the mesh was fine enough near the area in contact with the indenter. To keep the simulations computationally tractable, the mesh density was progressively coarsened away from the grain center. To avoid grain-boundary effects, the bottom and surrounding surfaces the specimen were constrained. The indenter was only allowed to move in the z direction. The FE mesh is illustrated in figure 2.

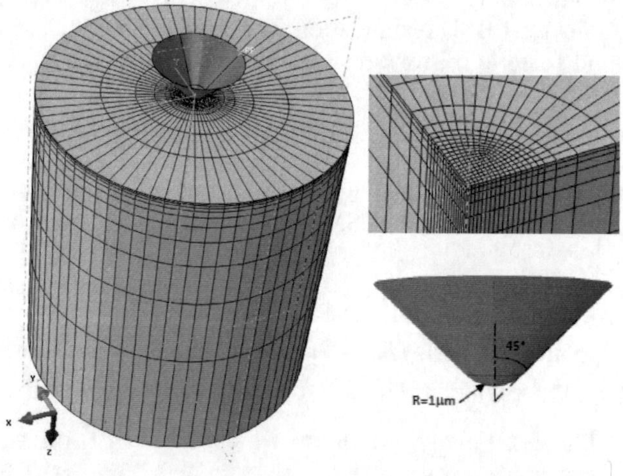

Figure 2. Three-dimensional Finite element mesh.

To investigate the sensitivity and confirm the convergence of the proposed mesh size on the numerical simulation, we have used the model of 13860 elements and the mesh size with the smallest element size of 60 nm as a benchmark. Three modified cases of mesh size were considered, namely: 10% smaller, 10% larger, and 30%. The obtained results showed that for the case with 30% larger element size (the smallest element size is 78 nm), the deviation of the penetration depth and the pile-up height from the benchmark was 5.2%. For the case with 10% larger element size, the deviation was 4.1% and for the case with 10% smaller size (the smallest element size is 54 nm) the deviation was only 0.4%. Based on this sensitivity analysis the selected model with 13860 elements is enough to get the converging solution.

The cono-spherical indenter was modelled as a fully rigid body, with a tip radius of 1 μm and a half included angle of 45°. A friction coefficient of 0.1 was assumed between the indenter and the zinc substrate. Even if friction is known to play a minor role on the load vs. displacement response during

IOP Conf. Series: Materials Science and Engineering **540** (2019) 012011 doi:10.1088/1757-899X/540/1/012011

indentation [18-19], the use of a friction coefficient enhances convergence by limiting mesh elements distortion in the contact area.
Furthermore, like experimental conditions, all simulations were performed under quasi-static, isothermal load-controlled conditions. In addition, nonlinearity option was used to account for both material and geometric (finite strains and rotations) nonlinearities.
The stiffness tensor **C** was considered to model the single crystal elastic response, which is composed with the six following terms: $C_{11}= C_{22}=$ 165 GPa, $C_{12}=$ 31.1 GPa, $C_{13}= C_{23}=$ 50 GPa, $C_{33}=$ 61.8 GPa, $C_{44}= C_{55}=$ 39.6 GPa, $C_{66}=$ 66.95 GPa.
Since plastic deformation is obtained by crystallographic dislocation slips, a total of 12 slip systems was considered: basal slip systems (3), prismatic slip systems (3) and $\Pi_2 \langle c+a \rangle$ slip systems. According to the proposed crystal plasticity model, four material parameters ($h_0^{(\alpha)}, \tau_0^{(\alpha)}, \tau_s^{(\alpha)}, \dot{\gamma}_0^{(\alpha)}$) should be identified for each family of slip systems.
To keep the inverse identification problem more tractable, we assumed the same reference shear strain rate on each slip system α $\left(\dot{\gamma}_0{}^{(\alpha)} = \dot{\gamma}_0 = 1.0\ 10^{-3}\right)$ and strain rate sensitivity exponent ($n = 45$).
Furthermore, the basal glide being the easy slip system for zinc, we assumed a constant ratio between basal and prismatic glide initial yield stresses $\left(\tau_0^{Prism}/\tau_0^{Basal} = 15\right)$, and between basal and pyramidal glide initial yield stresses $\left(\tau_0^{\pi\langle c+a\rangle}/\tau_0^{Basal} = 10\right)$, as suggested by Philippe *et al.* [20] and Fundenberger *et al.* [21]. Finally, the number of material parameters to identify is reduced to 7 ($h_0^{Basal}, h_0^{Prism}, h_0^{\pi\langle c+a\rangle}, \tau_0^{Basal}, \tau_s^{Basal}, \tau_s^{Prism}, \tau_s^{\pi<c+a>}$).
Those material parameters are obtained using an inverse identification procedure, by iteratively adjusting FE simulation and experimental load-penetration depth curves. The genetic algorithm MOGA-II to minimize the objective function defined as:

$$F_{obj} = \frac{1}{N}\sum_{i=1}^{N}\left(h_{num}(\mathrm{P},\mathrm{t}_i) - h_{exp}(t_i)\right)^2 \tag{5}$$

where P stands for the vector of unknown parameters, N is the number of data set, t_i denotes the time of the corresponding experimental point i, and $h_{num}(P,t_i)$ and $h_{exp}(t_i)$ are the indenter displacements numerically computed and experimentally measured, respectively.

5. Results and discussion

5.1. Identification of the crystal plasticity model parameters

To estimate the effect of cold rolling on the nanoindentation response of pure zinc grains, we have selected Grain #17 (AR: as-received annealed cold-rolled) and Grain #1 (TR: thickness reduction rate of 50%) for the identification of the crystal plasticity model parameters.
The identified parameters are summarized in table 3. It is found that increased percentage of cold reduction increases the critical resolved shear stresses (CRSS) $\tau_0^{(\alpha)}$. The main reasons for CRSS reduction are work hardening and grain refinement due to thickness reduction. Figure 3 depicts the experimental and numerical load-penetration depth curves issued from the identification procedure for the two selected grains. A good agreement between the experimental data and the FE simulation results can be observed for both selected grains. Moreover, the penetration depth at same load (1000 μN) is lower for Grain #1 with thickness reduction rate of 50% which is in accordance with the increase of the critical resolved shear stresses. Figure 4 shows the activity of slip systems in the area underneath the indenter. The basal and pyramidal systems are the more active systems. Basal system is the more active system for Grain #17 (figure 4a), while pyramidal system is more active for Grain #1 (figure 4c).

Table 3. Crystal plasticity model parameters identified.

Grain number - state	h_0^{Basal} (MPa)	h_0^{Pris} (MPa)	$h_0^{\pi\langle c+a\rangle}$ (MPa)	τ_0^{Basal} (MPa)	τ_s^{Basal} (MPa)	τ_s^{Pris} (MPa)	$\tau_s^{\pi<c+a>}$ (MPa)
1 - TR	214	635	1065	4.3	236	230	44
17 - AR	146	994	1364	3.49	284.5	136	43

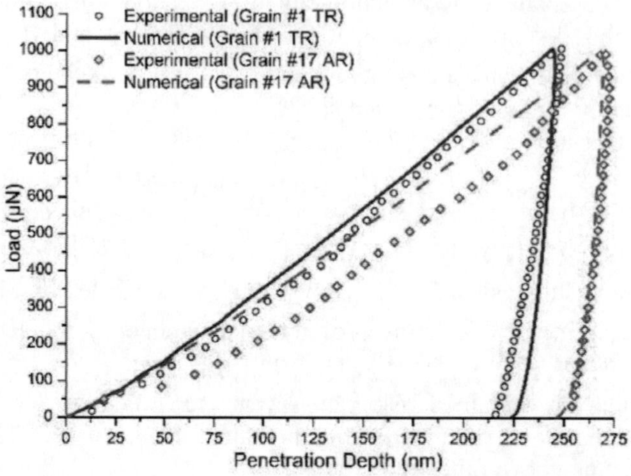

Figure 3. Experimental and simulated load-penetration depth curves.

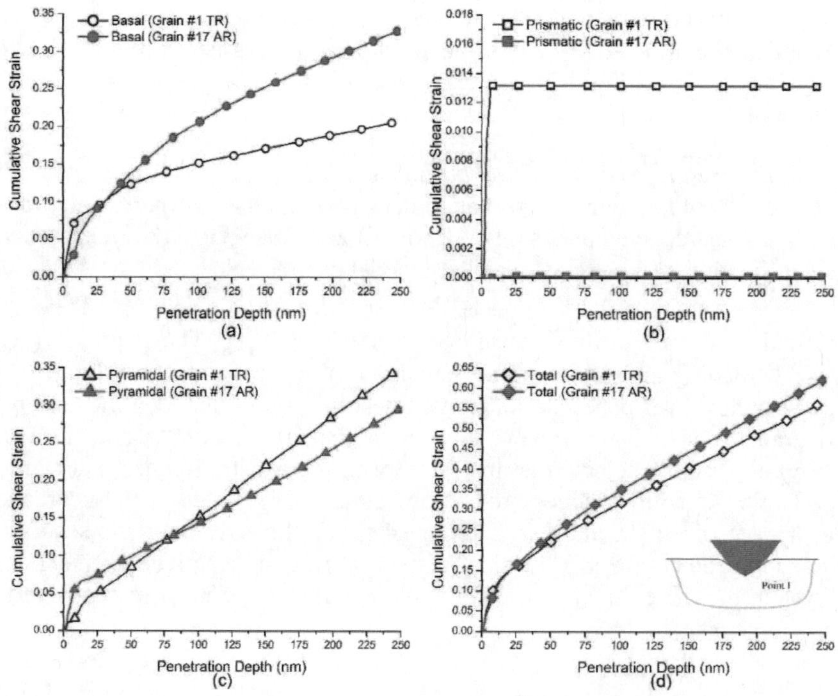

Figure 4. Evolution of $\gamma^{(\alpha)}$ in the area underneath the indenter.

5.2. Validation of the crystal plasticity model

To verify the validity of the proposed model, indentation tests performed on grains of different crystalline orientations were simulated using the CPFEM model with the identified parameters given in table 3. Grain #15 (AR: as-received annealed cold-rolled) and Grain #15 (TR: thickness reduction rate of 50%) were selected given their quite distinct location in the standard stereographic triangle, involving thus different slip systems activity rates.

The comparison between experimental data and numerical simulations is showed in figure 5. It is worth noting that the simulation results for both grains are in good agreement with the experimental data. Those results validate the identified model for a wide range of orientations.

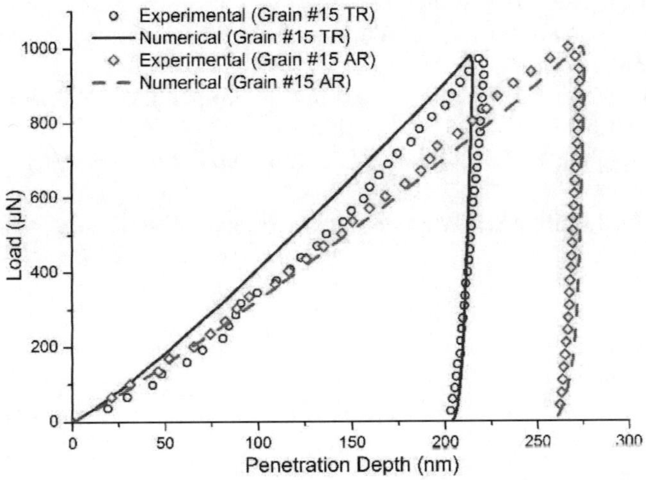

Figure 5. Experimental and simulated load-penetration depth curves.

6. Conclusion

In this work, we studied the orientation-dependent response of pure zinc grains subjected to two cold-rolling conditions. Instrumented indentation experiments were performed on grains of different crystalline orientations for as-received annealed cold-rolled material and material with thickness reduction rate of 50%. A crystal plasticity model was implemented in the FE code Abaqus using a user subroutine UMAT. The model parameters for both materials were identified by solving an inverse problem. It was found that increased percentage of cold reduction increases the critical resolved shear stresses (CRSS). Experimental data and numerical results are in good agreement for a wide range of orientations.

7. References

[1] H Takuda, T Yoshii and N Hatta 1999 *J. Mater. Process. Techn.* **89–90** p 135-140.
[2] T Kuwabara, C Katami, M Kikuchi, T Shindo and T Ohwue 2001 *In: Proceedings of the Seventh International Conference on Numerical Methods in Industrial Forming Processes, Toyohashi, Japan, 18-20 June,* p 781
[3] O Cazacu and F Barlat 2004 *Int. J Plast.* **20** p 2027-2045
[4] O Cazacu, B Plunkett and F Barlat 2006 *Int. J. Plast.* **22** p 1171-1194
[5] B Plunkett, R A Lebensohn, O Cazacu and F Barlat 2006 *Acta Mater.* **54** p 4159-4169
[6] M E Nixon, O Cazacu and R A Lebensohn 2010 *Int. J. Plast.* **26** p 516-532
[7] S K Khan, S Yu and H Liu 2012 *Int. J. Plast.* **38** p 14-26
[8] J H Yoon, O Cazacu and R K Mishra 2014 *Mater. Sci. Eng. A* **565** p 203-212

[9] C Zambaldi, Y Yang, T R Bieler and D Raabe 2012 *Journal of Materials Research* **27** p 356-367
[10] L Cauvin, B Raghavan, S Bouvier, X Wang and F Meraghni 2018 *Mater. Sci. Eng. A* **729** p 106-118
[11] N P T Nguyen, F Abbès, B Abbès and Y Li 2018 *H Nguyen-Xuan et al. (eds.), Proceedings of the International Conference on Advances in Computational Mechanics 2007, Lecture Notes in Mechanical Engineering* p 157-169
[12] D Peirce, R J Asaro and A Needleman 1983 *Acta Metall.* **31** p 1951-1976
[13] G I Taylor 1938 *J. Inst. Metals* **62** p 307-324
[14] M H Yoo and C T Wei 1966 *Philosophical Magazine* **13** p 759-775
[15] M H Yoo and J K Lee 1991 *Philosophical Magazine* **63** p 987-1000
[16] Y Huang 1991 *Mech Report 178* (Harvard University)
[17] J W Kysar and P Hall 1997 (Harvard University)
[18] Y Liu, B Wang, M Yoshino, S Roy, H Lu and R Komanduri 2005 *Journal of the Mechanics and Physics of Solids* **53** p 2718–2741
[19] Y Liu, S Varghese, J Ma, M Yoshino, H Lu and R Komanduri 2008 *International Journal of Plasticity* **24** p 1990–2015
[20] M J Philippe, M Serghat, P V Houtte and C Esling 1994 *Acta Metallurgica & Materalia* **42** p 239-250
[21] J J Fundenberger, M J Philippe, F Wagner and C Esling 1997 *Acta Materialia* **45** p 4041-4055

IOP Conf. Series: Materials Science and Engineering **540** (2019) 012012 doi:10.1088/1757-899X/540/1/012012

Springback Analysis in Flexible-bending Process of Tubes

Chaofan Shi[1,2], **Wenzhi Fu**[1,2*] , **Mingzhe Li**[1,2]**and Zhuo Yi**[1,2]

1College of Materials Science and Engineering, Jilin University,Nanling Campus, 5988 Renmin Street, Changchun130025, People's Republic of China
2Roll Forging Research Institute, Jilin University, Nanling Campus, 5988 Renmin Street, Changchun 130025, People's Republic of China
*E-mail: fwz@jlu.edu.cn

Abstract. Flexible-bending, as a new bending technology, is widely used in the processing of profiles and tubes, especially in the production of small batches and multi-curvature. In flexible-bending process of tubes, the stress is complex and the springback problem exists which affects the forming accuracy and quality of bent tubes. In this paper, the springback is studied and solved by springback compensation by means of finite element numerical simulation and experimental verification. The results show that the wall thickness of tubes and the bending radii have important influence on springback. Springback compensation can effectively solve the problem of forming error caused by springback because that the forming radii after compensation can approach the target radius gradually. The experiments were carried out with flexible-bending equipment. The simulation results were in agreement with the experiments, which proved the reliability of the simulation.

1. Introduction

Because of the advantages of light weight and high strength, the demand and research of bent tubes are increasing. Traditional methods of bending tubes, such as rotary-draw-bending, bend tubes according to the shape of the die, which greatly affects the processing efficiency and production cost. Compared with traditional bending forming, flexible-bending has the advantage of processing different forming radii without changing the die. Murata [1-2] first proposed the flexible-bending method called MOS and designed the MOS equipment of bending tubes. Li Pengfei et al. [3-4] studied the flexible-bending of tubes and profiles of continuous varying curvature. Zhou Yongping [5-6] analyzed the elimination of forming defects of profiles and the residual stress of bent tubes in flexible-bending process.

The process of flexible-bending of tubes involves material nonlinearity, geometric nonlinearity and boundary-condition nonlinearity. The stress state and deformation situation are complex, which often cause forming defects such as springback. In this paper, the springback defect is studied by numerical simulation and experimental verification and the corresponding solutions are put forward.

2. The principle of flexible-bending

Figure 1 shows the principle of flexible-bending. The forming device mainly consists of a pusher, a guiding device, a bend die and a bearing. Among them, the guiding device is fixed on the workbench, the pusher feeds the tube along the workbench direction (Z direction), the bearing can move on the plane (XY plane) vertical to the workbench to drive the offset of the die, the bend die is embedded in the bearing which can move along with the bearing and rotate at a certain angle under the action of the tube and bearing. Before the bending, adjust their centers so that they are on the same horizontal line and then the tube is penetrated into the guide and die. In the bending process, move the bearing to make the die shift some distance and rotate at a certain angle, and then feed the tube constantly to complete the continuous forming. The forming radius of the tube is determined by the die offset and different bending radii can be realized with different offset, so we can achieve the flexible-bending.

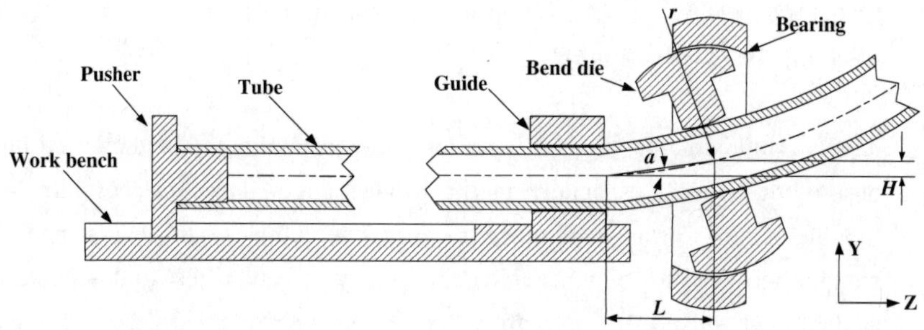

Figure 1. Schematic diagram of flexible-bending.

In the process of flexible-bending of tubes, the horizontal distance between the right side of the guiding device and the geometric center of the bending die is *L*, the offset of the bending die along the Y direction is *H*. The theoretical bending radius is *r* and the theoretical bending angle is *a*.

According to the geometric relationship in the diagram, the *r* and *a* can be obtained

$$r = \frac{L^2 + H^2}{2H} \tag{1}$$

$$a = \arctan\frac{H}{L} \tag{2}$$

From the formulas and related geometric relationships, It can be seen that when *H* increases keeping *L* unchanged or *L* decreases keeping *H* unchanged, the theoretical bending angle *a* increases, which means that the corresponding theoretical forming radius *r* decreases.

3. Finite element model

In this paper, ABAQUS/explicit is used to simulate the process of flexible-bending. Figure 2 shows the finite element model. In order to facilitate calculation, the bearing, the bend die, the guide and the pusher are set as rigid bodies divided into R3D4 rigid body elements; the tube is set as a deformable body with the diameter of 12 mm and the thickness of 1 mm and is divided into C3D8R solid elements in the simulation. The material of the tube is 304 stainless steel and the related material parameters are as follows: density is 7930 Kg/m^3, elastic modulus is 204 Gpa, yield strength is 205 Mpa and Poisson's ratio is 0.25.Considering friction, the friction coefficient at contact surfaces is 0.2.

IOP Conf. Series: Materials Science and Engineering **540** (2019) 012012 doi:10.1088/1757-899X/540/1/012012

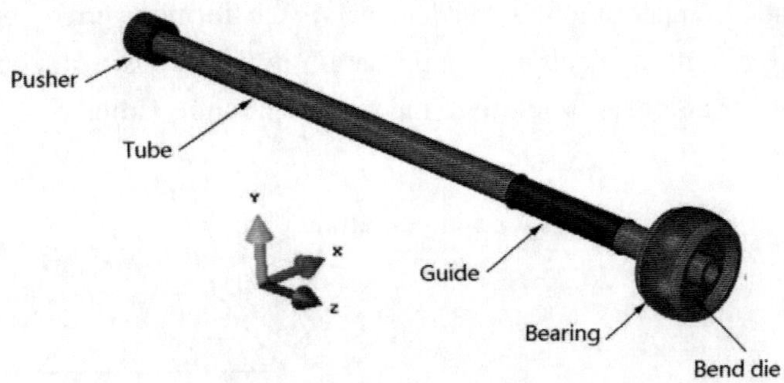

Figure 2. Finite element model.

4. Analysis of numerical simulation results

4.1 Analysis of springback

In order to explore the influence of wall thickness and bending radii on springback, we simulated the bending effect of different wall thickness and different bending radii. For facilitating the analysis, we use the radii difference (Δr) of before and after springback to express the springback amount.

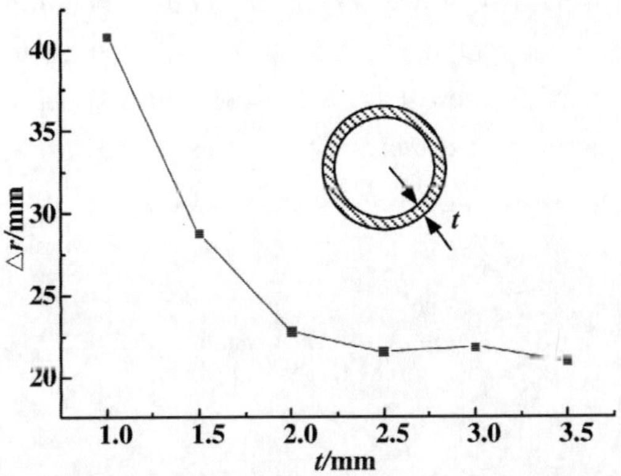

Figure 3. The influence of wall thickness on Springback.

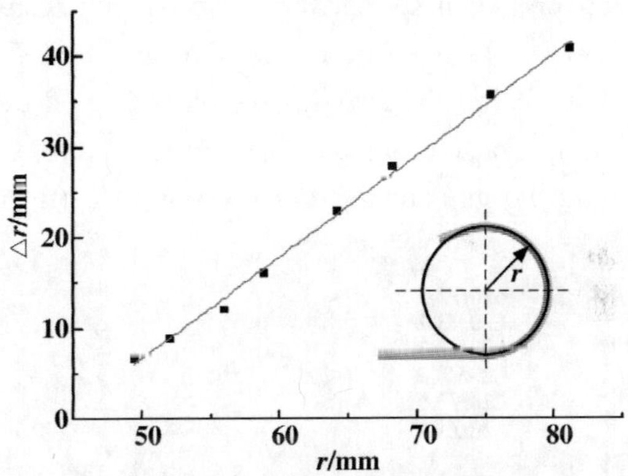

Figure 4. The influence of bending radii on Springback.

Figure 3 shows the changing tendency of Δr along with the wall thickness when the diameter D = 12 mm and theoretical bending radius r = 81.25 mm. It can be seen that Δr decreases rapidly and then tends to be stable with the increase of wall thickness. This is because the tube body volume increases and the proportion of plastic deformation increases, which weakens the springback. As shown in Figure 4, the springback under different bending radii is obtained when the outer diameter D = 12 mm and the wall thickness t = 1 mm. The radii difference Δr increases with the increase of bending radii and they show a linear relationship. This is because that the effect of elastic deformation increases with the increase of bending radii, which makes the springback phenomenon obvious.

4.2 Springback compensation

IOP Conf. Series: Materials Science and Engineering **540** (2019) 012012 doi:10.1088/1757-899X/540/1/012012

In this paper, springback compensation is used to solve the forming error caused by springback. Springback compensation is giving bending compensation in the opposite direction of springback so that the forming radius after unloading is close to the target curvature radius.

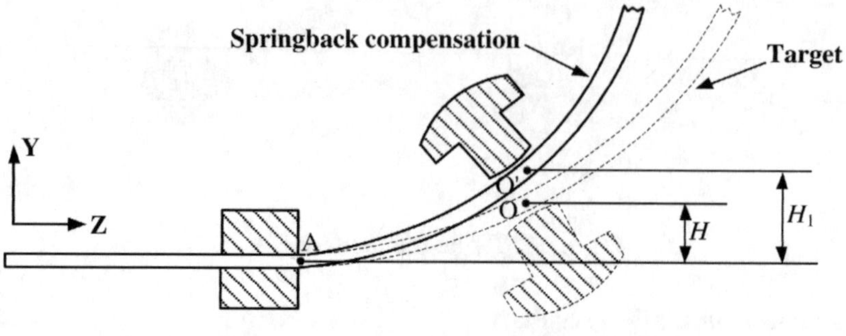

Figure 5. Schematic diagram of springback compensation.

Figure 5 is a diagram of springback compensation in flexible-bending process. O and *H* represent the center of the bending die and the offset of the die respectively when forming the target radius according to the theoretical formula; O'and H_1 are the corresponding position and the offset of the die respectively when springback compensation is applied. In this paper, we set the theoretical bending radius (r=81.25 mm) of L= 20 mm and H=2.5 mm as the target curvature radius. According to formulas, the bending compensation is carried out by increasing the offset of the die while keeping L=20 mm. Considering Δr when r=81.25 mm, take the increase of die offset (ΔH) as 0.6, 0.7, 0.8, 0.9, 1 and 1.1 mm and the corresponding die offset (H) is 3.1, 3.2, 3.3, 3.4, 3.5 and 3.6 mm.

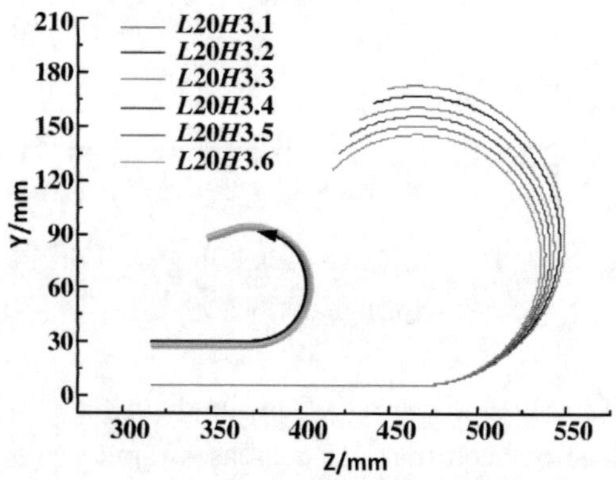

Figure 6. The node shape of outer surface of the inside of bent tubes.

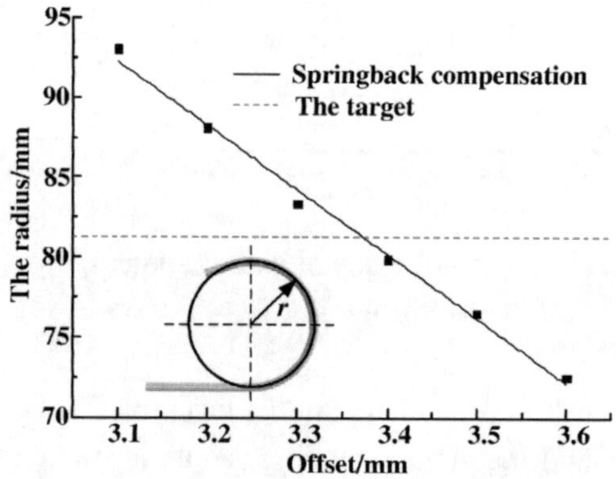

Figure 7. The bending radii of the neutral layer.

The numerical simulations were calculated and the bent tubes after springback compensation were obtained. The nodes were extracted from the outer surface of the inside of bent tubes, and the radii of the neutral layer of the bent tubes were measured and calculated. Figure 6 shows the node shape of outer surface of the inside of bent tubes .Figure 7 shows the radii of the neutral layer under different bending compensation. It can be seen that with the increase of the offset of the die, the forming radius decreases and the corresponding bending compensation increases. When *H* is smaller meaning less bending compensation, the forming radius after unloading is larger than the target curvature radius; when *H* is larger meaning more compensation, the forming radius after unloading is smaller than the target curvature radius. This phenomenon makes the forming radius approach the target radius continuously and then the gap becomes larger. At the same time, it shows that there is a value of *H* which can make the bent tube produce a suitable bending compensation and make the forming radius coincide with the target curvature radius.

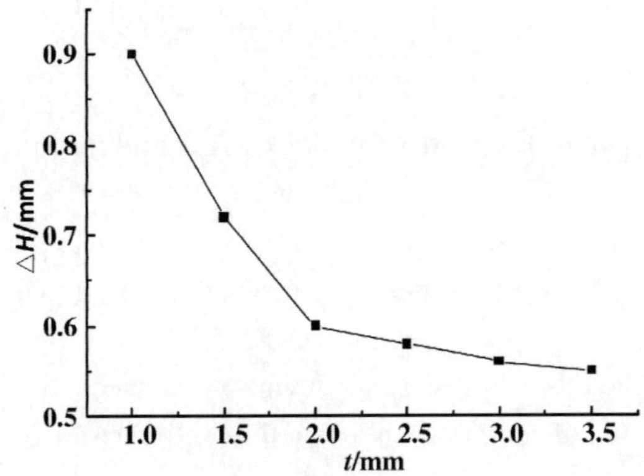

Figure 8. Springback compensation under different wall thickness.

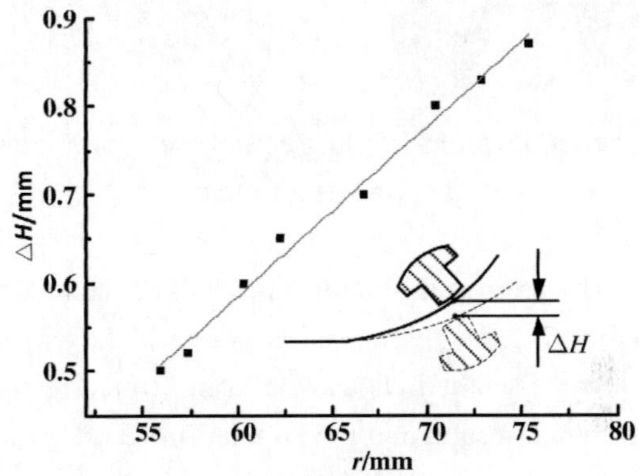

Figure 9. Springback compensation under different bending radii.

Because we adjust the bending degree by increasing *H*, so the amount of springback compensation is expressed by ΔH . The springback compensation of different wall thickness with diameter D=12 mm and bending radius r=81.25 mm was studied. The ΔH under different wall thickness is shown in the figure 8. It can be seen that with the increase of wall thickness, the ΔH decreases rapidly and then tends to be gentle. This is because springback amount decreases with the increase of wall thickness and the corresponding bending compensation shows the same trend. The springback compensation of different bending radii was studied when diameter D = 12 mm, wall thickness t = 1 mm and L = 20 mm. Figure 9 shows the springback compensation under different bending radii. It can be seen that with the increase of bending radii, the ΔH increases and there tends to be a linear relationship between them. This is because the larger the bending radius is, the larger the springback amount is and the corresponding bent tubes need more bending compensation.

5. Experiment

In this paper, the flexible-bending forming equipment developed by Jilin University independently was used for experiments. Figure 10(a) and 10(b) are flexible-bending equipment and experimental results of springback compensation respectively.

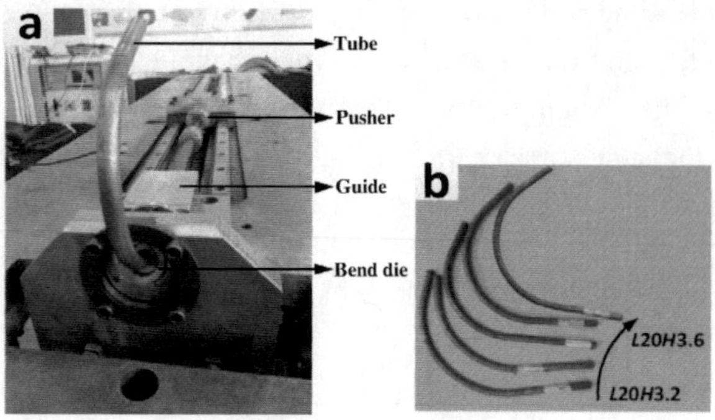

Figure 10. Flexible-bending equipment and related experiment results.

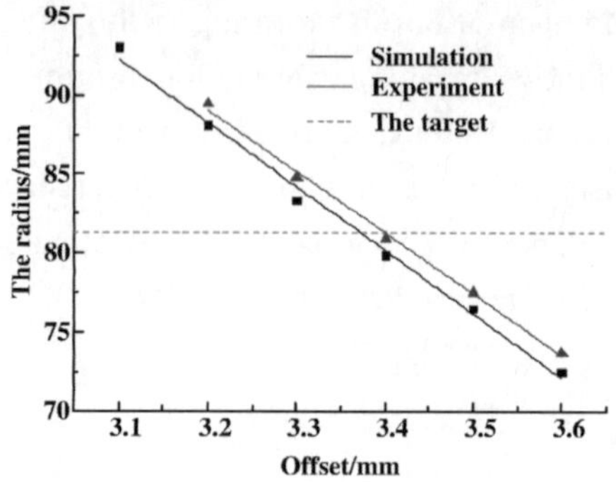

Figure 11. Comparison of experimental results and the target.

After the experiment, the forming radii of the neutral layer of the bent tubes were measured. As shown in Figure 11, the relationships between the experimental results, simulation results and the target radius are discussed. It can be seen from the figure that there is a certain deviation between the experimental results and the simulation results, which is due to factors such as friction and clearance. The experimental results are in agreement with the simulation results and also show a trend of approaching the target radius and then the gap becomes larger. It shows that bending compensation in the opposite direction of springback can make the forming radii approach the target curvature radius and effectively solve the problem of forming error caused by springback.

6. Conclusion

In the process of flexible-bending, there is forming springback. We studied it and put forward the method of springback compensation to solve it.

(1)There is springback defect for 304 stainless steel tubes in flexible-bending process. The wall thickness and bending radii have important influence on springback. With the increase of wall thickness, the springback gradually decreases and then tends to be stable. The springback increases linearly with the increase of bending radii.

(2) By means of springback compensation, the forming radii can be close to the target radius, which can effectively reduce the forming error caused by springback and improve the forming accuracy. The experimental results validate the effectiveness of the method of and the reliability of the simulation.

(3) With the increase of wall thickness, the amount of springback compensation ΔH decreases rapidly and then the trend becomes gentle. With the increase of bending radii, the amount of springback compensation ΔH increases linearly.

References

[1] Murata M, Ohashi N and Suzuki H 1989 New flexible penetration bending of a tube : 1st Report, a study of MOS bending method. *Transactions of the Japan Society of Mechanical Engineers C*, **55**:2488-2492.

[2] Murata M 1996 Effects of inclination of die and material of circular tube in MOS bending method. *Transactions of the Japan Society of Mechanical Engineers C*, **62(601)**:3669-3675.

[3] Li P, Wang L and Li M 2016 Flexible-bending of profiles with asymmetric cross-section and elimination of side bending defect. *International Journal of Advanced Manufacturing Technology*, **87(9-12)**:1-7.

[4] Li P, Wang L and Li M 2016 Flexible-bending of profiles and tubes of continuous varying radii. *International Journal of Advanced Manufacturing Technology*, **88(5-8)**:1-7.

[5] Zhou Y, Li P and Li M 2017 Application and correction of L-shaped thin-wall aluminum in flexible-bending processing. *International Journal of Advanced Manufacturing Technology*, **92(1)**:1-8.

[6] Zhou Y, Li P and Li M 2017 Residual stress and springback analysis for 304 stainless steel tubes in flexible-bending process. *International Journal of Advanced Manufacturing Technology*, **94(1)**:1-9.

IOP Conf. Series: Materials Science and Engineering **540** (2019) 012013 doi:10.1088/1757-899X/540/1/012013

Study of Surface Roughness of Electronic Substrate on Abrasive Belt Grinding

Q.B. Tao[1], M.T. Tran[2] and H.T. Bui[3]

1 Mechanical Engineering Department, University of Science and Technology, The University of Danang, VietNam
2 Institute of Mechanical Technology and Automation, Mechanical Engineering Department, University of Science and Technology, The University of Danang, VietNam
3 Mechanical Engineering Department, University of Technology and Education, The University of Danang, VietNam
E-mail: tqbang@dut.udn.vn; tmthong@dut.udn.vn; bhthong@ute.udn.vn

Abstract. Machining processing industries have continuously developed and improved technologies and processes to transform finished product to obtain better super finished product quality and thus increase products. Abrasive machining is one of the most important of these processes and therefore merits special attention and study. Indeed, grinding is the process of removing metal by the application of abrasives which are bonded to form a rotating wheel or belt. When the moving abrasive particles contact the workpiece, they act as tiny cutting tools, each particle cutting a tiny chip from the workpiece. The abrasive belt grinding is efficient, economic, widely used and being said "universal grinding". It can get high machining accuracy and surface quality. Surface quality is a very important aspect of machining quality and it is the most important parameter to measure the surface quality. Many factors affect the surface roughness such as performance of abrasive belt, the amount of abrasive belt grinding, hardness of contact wheel. And the most important one is the amount of grinding. For electronic packaging, before soldering, copper substrates have to be polished by both chemical and mechanical polishing. Therefore, this article is aimed at that status so that to design and manufacture a new abrasive belt grinding machine. And then the surfaces for soldering of Cu substrates are ground and polished using the machine. Finally, the surface roughness of copper substrates polished by the abrasive belt grinding machine is evaluated and compared with other grinding machines.

1. Introduction

In mechanical machining, grinding is the one of the most important metal machining methods which is recently applied for high-tech grinders. Grinding process is typically used in the last machining process that creating a flat and smooth surface through abrasion. The process can attain accurate level 6-7, high surface smoothness (surface roughness level 7-8 and higher) and good surface quality. Up to now, grinding machines still account up to 30% of total machine tools and up to 60% in accurate mechanical industry. The principle of grinding that uses abrasive particles to cut a tiny chip from workpiece surface. Each abrasive particle coated in belt can comprehend as cutting tools like milling and turning cutter. Grinding is the multistage marching which particularly consists of coarse grinding stage and fine grinding stage [1]. In the initial stage with coarse grinding, this is preliminary machining step, metal removal takes place significantly but surface quality and accuracy level are poor.

To improve all above cons, at the next stage, machine will be used some different finer emery paper grades. In last fine grinding stage, with detailed machining process, workpiece has high-smoothness and required accuracy. Therefore, grinding is often used in last machining step in wood and metal machining industry. Recently, bench grinder is widely used in technology.
The evolution of the surface roughness has been studied from the very moment the abrasive wheel is freshly dressed until it stabilizes. A close relationship between the measured roughness and the radial wear of the wheel has been found. Furthermore, these days, with the breakthrough development of technology, machines are required to work with high accuracy and efficiency performance. In order to reduce labors and gain higher productivity, researches of improvement and innovation need to be conducted to meet those requirements [2-5].
To increase the efficiency of grinding, abrasive belt grinding machines are increasingly used. Abrasive belt grinding machines are typed of machine tools using abrasive feature of abrasive belt to grind the surface detail. The superiority of abrasive belt grinding machines is being proved more and more clearly nowadays. In addition to the obsolescence of grinding machines using flap disc due to their low productivity, flap disc is also easily unsafely breakable, changeable and requires regular adjustment. Particularly for the need of coarse grinding, workpiece processing after laser, plasma or oxygen-gas cutting, abrasive belt grinding machines show excellently preeminent metal removable performance (when using coarse abrasion belt P36, P60). Furthermore, with the large belt and significant working range, the flexibility of grinding products using this machine increases considerably comparing to the traditional grinding. Especially, regarding the plants where manipulating arms are used to operate shaping abrasion, grind and chamfer edges of huge workpiece quantity, it is such an ideal choice to apply abrasive belt grinding machines because their parameters are not affected by the impact of worn-out or shape-changing status of emery belt surface. Therefore, the operating program is simple, which needs no sensor to recognize abrasive flap rockiness and flap-fixing structure. As a result, abrasive belt grinding machines have been widely used in mechanical engineering, wood processing, furniture manufacturing and handicrafts creating, etc. Application for fine grinding, metal coarse grinding, stainless steel.
In addition, surface quality is a very important aspect of machining quality. Surface roughness is the most important parameter to measure the surface quality [6, 7]. Many factors affect the surface roughness, such as performance of abrasive belt, the amount of abrasive belt grinding, hardness of contact wheel. And the most important one is the amount of grinding. Also, in electronic packaging, lap-shear specimens that are prepared by soldering two copper (Cu) substrates using solder materials. Surface for soldering of Cu is very important for making good joints. In order to evaluate shear, creep and thermal fatigue behavior of solder joints for different solder alloys, the lap-shear technique has been widely used. The Cu substrates are previously cut with electrical discharge machining (EDM) according to a specific geometry, and then the surfaces for soldering are ground and polished using the belt grinding machine. Therefore, in this study, first, it designs abrasive belt grinding device, then uses the device for polishing Copper substrates that used in electronic packaging, and finally, analyses the surface roughness of Copper substrates to evaluate the efficiency of the device.

2. Methodology and experimental procedure

2.1. Design of belt grinding device

The methods of using hand grinders and bench grinders are low-cost but the surface quality of products is not ideal. Moreover, these methods are prone to serious accidents due to broken flap disc as mentioned above. Therefore, abrasive belt grinding is superior to improve flexibility, surface quality and users' safety. Through the systematic approach to optimize parametric setting such as work material properties, the grit, abrasive belt type, belt speed, contact wheel hardness, grinding pressure and so forth to obtain the required output and precision in abrasive belt grinding. Currently, in the world, abrasive belt grinding machines are widely applied not only in mechanical engineering but also in construction engineering, health, hard-material processing, especially wood and furniture manufacturing industry. Many international companies have produced and sold some kinds of abrasive

IOP Conf. Series: Materials Science and Engineering **540** (2019) 012013 doi:10.1088/1757-899X/540/1/012013

belt grinding machines in the international market, however, in general, the function and flexibility of the products are not desirable; they are also high-cost.

As discussed above, the objective of this study is to design and produce a multipurpose abrasive belt grinding machine with low costs but high precision and efficiency and then the device is used for polishing Copper substrates in electronic packaging. To meet these requirements, by changing the rotating direction of abrasive belt on grinding bracket, the machine is used as a versatile grinder. Firstly, in order to save time, human input and also fabricating materials, it is necessary to use Solidworks software which can be designing and simulating overall mechanism. This software helps us to presume machine structure before manufacturing to consider dimensions and optimization that is compatible to design requirements. The overall model of belt grinding machine is shown in figure 1. All parts of the device are building in 3D space with accuracy dimensions.

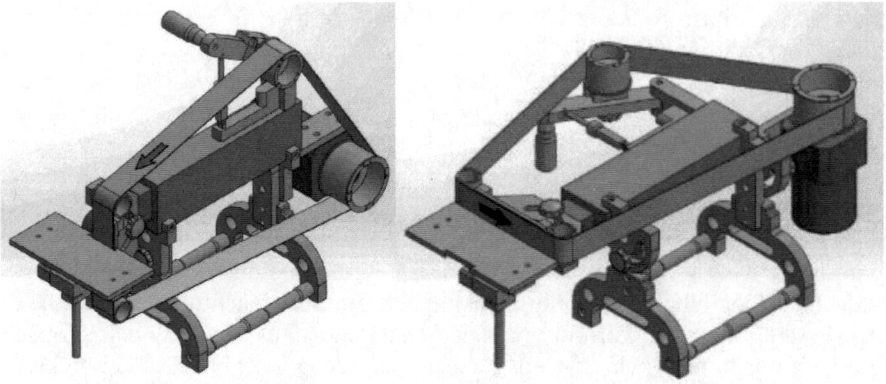

Figure 1. Belt grinding model.

The overall machine is modeling after using simulating software. The motor used in this machine is 3 phase AC motor because of low-price, simple operation and convenient maintenance. The motor shalf is mounted directly to roller. The specifications of motor are below:

- **MOTOR:** Yaskawa Electric
- Type: Induction motor
- Phase: 3, Volts: 200, Amps: 3.5/3.2 amp, HP: 1, RPM: 1420/1700, Hertz: 50/60, Poles: 4 Rating: Continuous
- Bearing No: 6205 ZZA (Max. RPM 18,000)

Roller is connected to motor shalf directly, when rotating motor shalf, this roller transmits power to other rollers through abrasive belt. To decrease the weight of machine, rollers is fabricated by aluminum material with diameter parameters alternatively which is D100, D60, 2*D40 (mm). Besides, the remaining parts of machine is fabrication by pipe and rectangular steel such as frame, tensioning part, grinding bracket.

For controller system, the controlling diagram of grinding machine is shown in figure 2. Variable frequency drive Schneider ATV12H075M2 is chosen to adjust 3 phase AC Motor speed. Up to now, VFD is applied for many machines in diverse fields, especially for industry and civil. The application of VFD is simple motor speed adjusting, low vibration boot, energy saving, overloaded motor protecting and high-performance increasing. VFD allows to adjust motor speed with the turn of knob(potentiometer). It also has an on/off power switch, start/stop switch and led signal light. Especially, it's compatible with a forward/reverse switch as an upgrade.

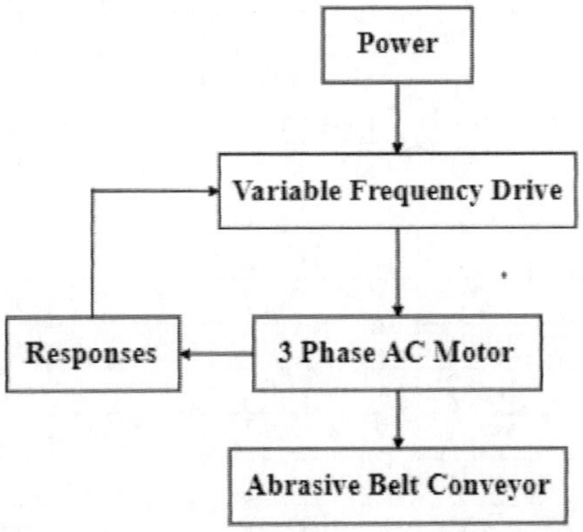

Figure 2. Grinding machine controlling system diagram.

2.2. Copper substrates for electronic packaging

In the study, the grinding device is used to polishing Copper substrates that used for lap-shear experiment as shown in figure 3. The lap-shear technique has been widely used to evaluate shear, creep and thermal fatigue behavior of solder joints for different solder alloys. A lap shear specimen is prepared by soldering two Cu substrates using solder materials. The Cu substrates are previously cut with EDM according to a specific geometry, and then the surfaces for soldering are ground and polished using the belt grinding machine. Finally, the impurities on the Cu substrates are removed by dipping them into 50% nitric acid for 20s and then quickly in acetone.

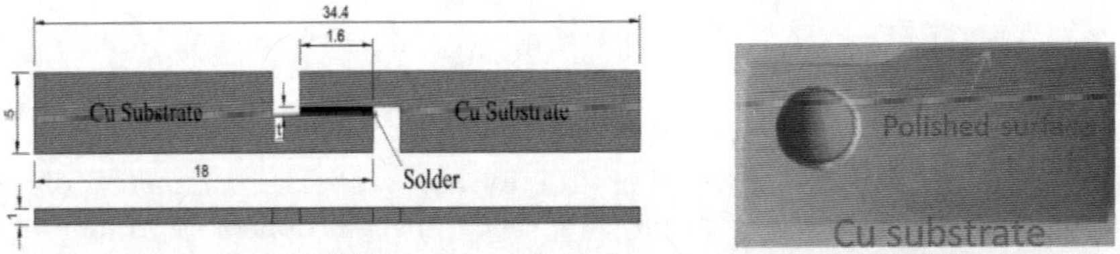

(a) (b)

Figure 3. The lap-shear joint specimen and Cu substrate for polishing.

3. Results and discussion

This research takes into account the effect of some process parameters of surface grinding process i.e. belt speed (rpm), and depth of cut on surface roughness with reference to copper as workpiece material. All other process parameters of surface grinding operation are kept constant. Due to its wide usage as a substrate material in electronic packages and scientific experiments, copper was chosen as the substrates. Therefore, this analysis is done to study the effect of input parameters on surface finish of copper in surface grinding operation. A variable frequency drive is used to vary the rpm (belt speed) of the grinding belt to three different values. A total of 9 experiments are done on each material by varying belt speed (rpm) and feed rate (mm/r). Later surface roughness of each specimen is checked using Mitutoyo Surface roughness tester with cut-off length as 0.25cm. Table 1 shows the experimental results for surface roughness of copper material at different technological parameters.

Table1. The surface roughness of copper material after different technological parameters.

Exp. No.	Material	Speed n_w (rpm)	Depth of cut f (mm/r)	Surface Roughness Ra (μm)
1	Copper	1400	0.01	0.177
2	Copper	1400	0.02	0.185
3	Copper	1400	0.03	0.250
4	Copper	2000	0.01	0.145
5	Copper	2000	0.02	0.160
6	Copper	2000	0.03	0.173
7	Copper	2500	0.01	0.130
8	Copper	2500	0.02	0.147
9	Copper	2500	0.03	0.156

We can be seen from the table that with the improvement of workpiece rotate speed n_w, amount of metal into the cutting zone in unit time increases. Material removal rate increases, debris becomes thicken and grinding grooves increases, which leads to decrease of surface roughness. Indeed, when the speed increases from 1400 to 2000 (rpm) and the feed rate keeps constant at 0.01 (mm/r), the surface roughness decreases around 18% from 0.177 to 0.145 μm.
With increase in depth of cut, there is a proportional increase in the normal pressure at the point of contact of wheel and workpiece. Very less depth of cut will lead to less cutting and more rubbing due to small shear pressure or good toughness value of material. Large depth of cut will also lead to less cutting and more rubbing due to more wear, abnormal fracture and completely break-off of wheel grains. Thus, an optimum value needs to be determined depending upon material and machine/process parameters.

4. Conclusion

In this study, the belt grinding machine was designed and built. Then, some copper substrates that used in lap-shear joint specimen, were tested at different technological parameters of the machine to evaluate the surface roughness. The effects on work piece surface roughness brought by performance of abrasive belt work piece rotate speeds grinding depth and feed hardness of contact wheel and so on are studied in experiments. The results indicate that using the abrasive belt grinding can reduce the surface roughness of work pieces effectively and increase the efficiency of grinding process.

5. References

[1] Xiao G and Malkin S 1996 *CIRP Annals.* **45** 287
[2] Sung-San Cho, Yong-Kyoon Ryu, Seung-Young Lee 2002 *Inter. J. Mach. Tool. Manu.* **20** 229
[3] Ioan D Marinescu and W Brian Rowe 2013 *Tri. Abr. Mach.g Pro.* 243.
[4] M. J. Awhale, N. C. Chinchkar, V. P. Gunjawate, N. S. Phule, and A. V. Amrute 2015 *Inter. J. Res. Civ. Mech. Eng.* **02** 56
[5] Kandpal B C, Verma R K, Malhotra D, Kumar A 2012 *Inter. J. Emer. Tech. Adv. Eng.* **02** 21
[6] Huang Y 2009 *Mater. Sci. For.* **610** 915
[7] Junde Qi, Dinghua Zhang, Shan Li and Bing Chen 2016 *Proc. IMechE. J. Eng. Manu.* 10.

Acknowledgments

This work was supported by The University of Danang, University of Science and Technology, code number of Project: T2018-02-34.

IOP Conf. Series: Materials Science and Engineering **540** (2019) 012014 doi:10.1088/1757-899X/540/1/012014

Experimental Characterisation and Finite Element Modelling of Paperboard for the Design of Paperboard Packaging

V D Luong[1], B Abbès[1], F Abbès[1], J B Nolot[2] and D Erre[1,2]

1 GRESPI, University of Reims Champagne-Ardenne, France
2 ESIReims, University of Reims Champagne-Ardenne, France
E-mail: boussad.abbes@univ-reims.fr

Abstract. In this work, tensile tests were first conducted on the paperboard. Then, the in-plane orthotropic elastoplastic model (IPE-Isotropic Plasticity Equivalent model) was implemented in Abaqus software. The advantage of this model lies in its capability to describe the in-plane anisotropic behavior of paperboard by a limited number of material constants. The calibration procedure was handled by inverse identification method on tensile tests. The model was shown to predict satisfactorily in-plane stress-strain behavior for different orientations. Box compression tests (BCT) were conducted to investigate the performance of the implemented model. The predicted deformed boxes and force-displacement curves were compared to the experimental ones, displaying a good agreement.

1. Introduction

Paperboard is one of the most frequently used material for packaging due to its beneficial characteristics, such as low price, sustainability, and recycling. It is used to design containers to protect the product from hazards of the distribution, transportation and storage environment so that the product can be shipped to consumers without damage. The package performance relates the load during transportation or order to accurately protect the goods. It can also relate to the behavior during consumer contact. Mechanical tests, which offers prompt verification of structural strength in the product design stage for deciding whether proper buffering is required in the packing design, is one of crucial items for validating the design. Therefore, the package should be designed such that it performs well. In recent years, many authors have studied the mechanical properties of paperboards by using in-plane tensile tests, short compression tests, creasing tests, folding tests and bending tests [1-12,19]. These studies have shown that mechanical properties of the paperboards depends on many factors such as thickness of paperboard, the direction of bending (MD, CD). Thereby, it highlights the most important mechanical properties is bending stiffness, which is directly related to the mechanism of failure in paperboard. The results of these studies also indicated that crack length, crack propagation and delamination are the typical forms failure of paperboard. It is the main cause of the reduction of the mechanical properties of the paperboard when bending, folding and creasing.

Using finite element simulations allows avoiding numerous experimental tests and predicting possible failures during the early design stage [18]. However, to get results close to the experimental ones, we need to know precisely the mechanical behaviour of the material. Paperboard is an anisotropic material. In the process of converting paperboard into a functional container, the paperboard undergoes large deformations. Mäkelä and Östlund [13] proposed an orthotropic elastoplastic model for in plane stresses (IPE-Isotropic Plasticity Equivalent model) to accurately describe the behavior of paperboard. The parameters in the model are determined from simple tests. This model was implemented in Abaqus software using the user subroutine VUMAT [14]. Material parameters were determined by inverse method: finite element simulations of the tensile tests in MD, CD and 45° are used, and the parameters are varied systematically until agreement between simulations and experiments is achieved. To validate the identified model, we have developed a finite element model for the box compression test (BCT). The simulation results were then compared to the experimental ones.

2. Anisotropic Elastoplastic Behavior Model of Paperboard

Paperboard consists mainly of wood fibers with three components: cellulose, hemicellulose and lignin. The board used in the construction of a carton is an orthotropic material. The production of its constituent layers, the paper fibres adopt a preferred orientation, giving three characteristic directions: the machine direction (MD), the cross direction (CD), and out-of-plane direction (ZD). In this study, we used paperboard with a grammage of 320 g/m^2 and thickness is 0.51 mm

The in-plane properties of paperboard are relatively easy to determine by tensile tests. Due to its low thickness, out-of-plane properties are more difficult to obtain. Stenberg [15] showed that the Young's modulus along normal direction (ZD) is about 200 times lower than that of MD. Stenberg et al. [16] observed that the deformation in the plane is negligible during the compression according to the thickness. The Poisson coefficients ν_{xz} and ν_{yz} consequently are close to zero.

In this study, we used the IPE model [13] to describe the behavior of paperboard. The orthotropic elasticity behavior in plane stresses is defined by:

$$\{\sigma\} = \begin{Bmatrix} \sigma_x \\ \sigma_y \\ \sigma_{xy} \end{Bmatrix} = [H]\{\varepsilon^e\} = \frac{1}{(1-\nu_{xy}\nu_{yx})} \begin{bmatrix} E_x & \nu_{yx}E_x & 0 \\ \nu_{xy}E_y & E_y & 0 \\ 0 & 0 & G_{xy}(1-\nu_{xy}\nu_{yx}) \end{bmatrix} \begin{Bmatrix} \varepsilon_x^e \\ \varepsilon_y^e \\ \gamma_{xy}^e \end{Bmatrix} \tag{1}$$

The deviatoric stresses vector and the IPE plasticity criterion are given by:

$$\{s\} = \begin{Bmatrix} s_x \\ s_y \\ s_z \\ s_{xy} \end{Bmatrix} = [L]\{\sigma\} = \frac{1}{3} \begin{bmatrix} 2A & C-A-B & 0 \\ C-A-B & 2B & 0 \\ B-C-A & A-B-C & 0 \\ 0 & 0 & 3D \end{bmatrix} \begin{Bmatrix} \sigma_x \\ \sigma_y \\ \sigma_{xy} \end{Bmatrix} \tag{2}$$

$$f = \sigma_{eq} - Y = \left(\frac{3}{2}\langle s\rangle\{s\}\right)^{1/2} - E_0\left(\varepsilon_0 + \varepsilon_{eq}^p\right)^{1/n} = 0 \tag{3}$$

where Y is the yield stress, A, B, C, D, $E_0, \varepsilon_0, \varepsilon_{eq}^p, n$ are the parameters of the IPE model that can be determined using experimental tests.

This IPE model was implemented in the ABAQUS software using the VUMAT user subroutine.

3. Inverse identification of the anisotropic elastoplastic behavior of paperboard

Standard specimens were made using a cutting table (ZÜND M-1600). To ensure a better grip in the jaws when clamping the specimen, we glued pieces of rigid cardboard at both ends of the test specimens. Experimental tensile tests were carried out for three directions (MD, CD and 45°) at a crosshead speed of 10 mm/min under standard conditions (23°C and 50% relative humidity). The experimental tensile results are shown in figure 1.

To identify the parameters (A, B, C, D, E_0, ε_0, ε_{eq}^p, n), an inverse method was used: finite element simulations of the tensile tests in MD, CD and 45° are run, and the parameters are varied systematically until agreement between simulations and experiments is achieved. The objective function minimized is the quadratic difference between the numerical and the experimental tensile forces:

$$F_{obj} = \frac{1}{N}\sum_{i=1}^{N}\left(F_{num}(t_i) - F_{exp}(t_i)\right)^2 \tag{4}$$

where N is the number of data set, t_i denotes the time of the corresponding experimental point i and F_{num} and F_{exp} are the forces numerically computed and experimentally measured, respectively.

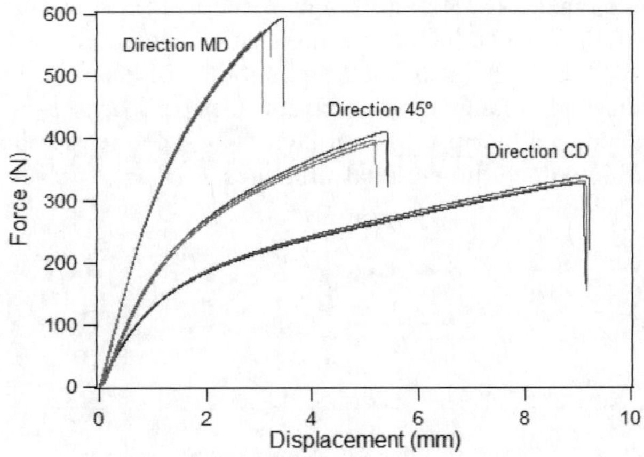

Figure 1. Force-displacement curves from tensile tests of paperboard.

In this study, we used the genetic optimization algorithm MOGA-II to minimize the objective function in equation (4). Figure 2 shows the experimental and numerical curves obtained by identification of the IPE model, and table 1 summarizes the parameters identified for paperboard.

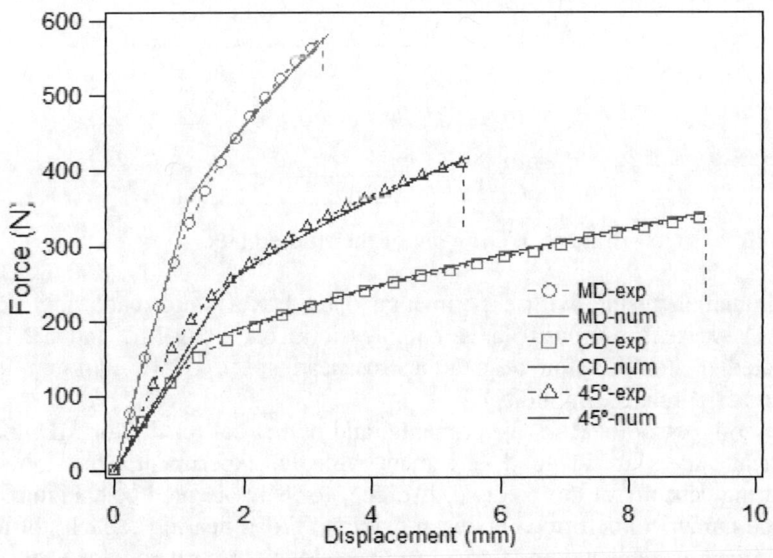

Figure 2. Experimental traction curves and identified for paperboard.

Table 1. Parameters of IPE model identified for paperboard.

E_x (MPa)	E_y (MPa)	ν_{xy}	G_{xy} (MPa)	E_0	n	A	B	C	D	ε_0
3710	1559	0.36	1200	301	2.31	1.0	2.04	2.48	1.08	0.0046

IOP Conf. Series: Materials Science and Engineering **540** (2019) 012014 doi:10.1088/1757-899X/540/1/012014

4. Finite element model of box compression test (BCT)

A finite element model of the box, whose drawing is shown in figure 3, was created using ABAQUS software. This model used to simulate compression behaviour of the box was meshed with reduced-integration four node shell elements (S4R) with 1 mm element size. The actual process of converting paperboard into a functional container includes a number of operations such as creasing to reduce resistance to bending and folding. It leads to the possibility of material damage such as loss of hardness, crack propagation and delamination of layers in paperboard [1-4, 6, 9]. The interlaminar bond strength is typically reduced by up to 80 per cent [17]. The creases then act as hinges, which encourage the board to fold into three-dimensional structures.

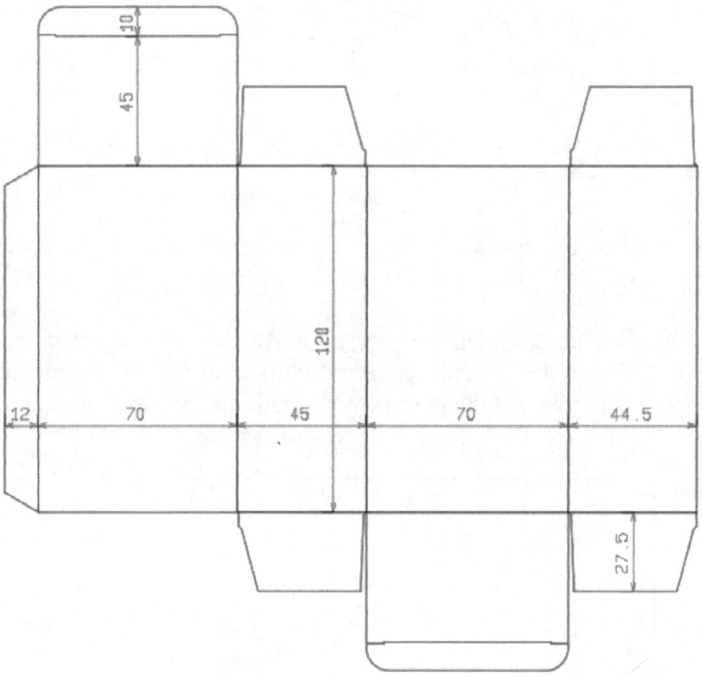

Figure 3. Drawing of the studied box.

To compare the simulation results with experimental ones, boxes were made using a cutting table (ZÜND M-1600). The boxes were cut to have compression axes along MD and CD directions. The boxes were subjected to compression tests at a crosshead speed of 10 mm/min under standard conditions (23°C and 50% relative humidity).

Figure 4 shows the comparison of BCT experimental and numerical results for MD and CD oriented boxes. The numerical curves are in good agreement with the experimental ones. As expected, the maximum load bearing capacity of the box (F_B) is higher for MD-oriented box. Figures 5 and 6 show the experimental and numerical deformed boxes at maximum load bearing capacity of the boxes (F_B). For both boxes, the load is concentrated in narrow zones along the corners. However, the stresses on the MD-oriented box (σ_{max} = 79.5 MPa) are higher than those on the CD-oriented box (σ_{max} = 59.0 MPa).

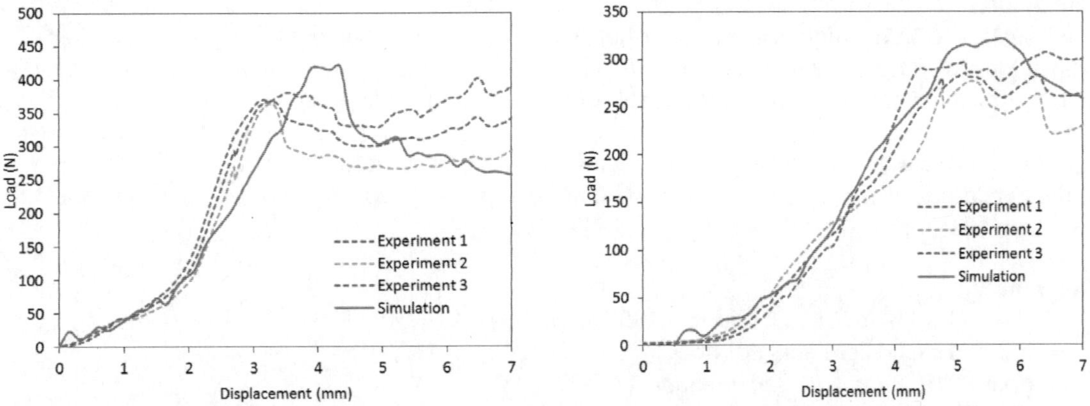

Figure 4. Comparison of BCT experimental and numerical results: MD (Left) - CD (Right).

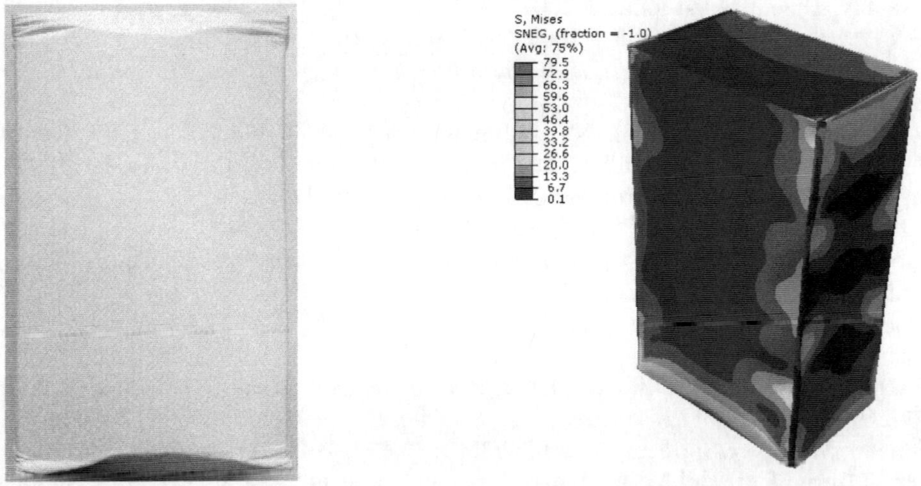

Figure 5. Comparison of experimental and numerical deformed MD-oriented box at F_B.

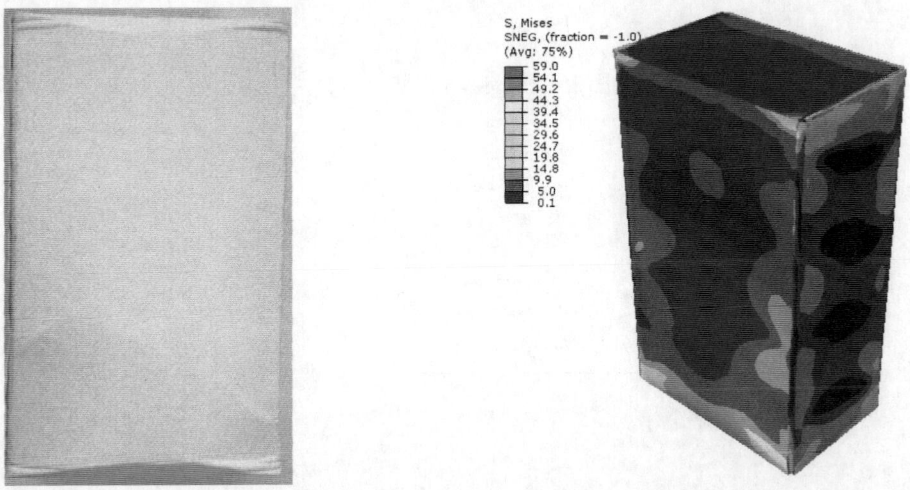

Figure 6. Comparison of experimental and numerical deformed CD-oriented box at F_B.

IOP Conf. Series: Materials Science and Engineering **540** (2019) 012014 doi:10.1088/1757-899X/540/1/012014

5. Conclusion

In this paper, we have implemented an orthotropic elastoplastic model for in plane stresses (IPE-Isotropic Plasticity Equivalent model) in Abaqus software using the user subroutine VUMAT. We have developed an inverse identification procedure by coupling the Abaqus finite element software and an optimization procedure to identify the material parameters of the paperboard. Box compression tests (BCT) were carried out to validate the identified model. The finite element model compares well with the experimental results in terms of maximum load bearing capacity and deformed shapes of the boxes.

6. References

[1] Carlsson L, Alf de R and Christer F 1983 *J. Mat. Sci.* **18** 1365-1373

[2] Carlsson L, Christer F and Alf de R 1980 *J. Mat. Sci.***15** 2636-2642

[3] Cavlin SI 1988 *Packag. Technol. Sci.* **1** 77-92

[4] Birger E 1998 *Packag. Technol. Sci.* **11** 131-140

[5] Giampieri A, Perego U and Borsari R 2011 *Int. J. Solids Struct.* **48** 2275-2287

[6] Cavlin S, Dunder I and Edholm B 1997 *Packag. Technol. Sci.* **10** 191-207

[7] Mikael N, Magnus J and Johan T 2009 *Int. J. Solids Struct.* **46** 2493-2505

[8] Beex L.A.A and Peerlings R.H.J 2009 *Int. J. Solids Struct.* **46** 4192-4207

[9] Juan C A 2012 *Stiffness Design of Paperboard Packages using the Finite Element Method* (Master of Science Thesis)

[10] Mentrasti L, Cannella F, Pupilli M and Jian S. D 2013 *Int. J. Solids Struct.* **50** 3089-3096

[11] Mentrasti L, Cannella F, Pupilli M and Jian S. D 2013 *Int. J. Solids Struct.* **50** 3097-3105

[12] Hui H, Hagman A and Nygårds M 2014 *Mech. Mat.* **69** 11-34

[13] Mäkelä P and Östlund S 2003 *Int. J. Solids Struct.* **40** 5599-5620

[14] ABAQUS 2016 *ABAQUS Documentation*, Dassault Systèmes, Providence, RI, USA

[15] Stenberg N 2003 *Int. J. Solids Struct* **40** 7483-7498.

[16] Stenberg N, Fellers C and Östlund S 2003, *J. Eng. Mat.Technol.* **123**(4) 184-190.

[17] Emslie AG and Brenneman RS 1967. *Tappi* **50** 289–297

[18] Luong V D, Abbès F, Abbès B, Duong P T M, Jean-Baptise N, Damien E and Ying-Qiao G 2017, *Proc. Int. Conf on Advances in Computational Mechanics 2017* (Phu Quoc Island, Vietnam: Singapore/Springer) p 369-380 DOI: 10.1007/978-981-10-7149-2_25

[19] Tobi F, Corné J. C, Tarl M.B, Alemayehu A and Umezuruike L.O 2018 *Bio. Engineer.* **174** 20-40

IOP Conf. Series: Materials Science and Engineering **540** (2019) 012015 doi:10.1088/1757-899X/540/1/012015

Modeling and Experimental of the Mechanical Behavior of Seat Cushion for Wheelchair Users

H-T. Bui [1], L-V. Nguyen[1], K. Debray[2], Q-B. Tao[3], P. Lestriez[2]

1 The University of Danang – University of Technology and Education, VietNam
2 URCA/GRESPI – University of Reims Champagne-Ardenne, France
3 The University of Danang – University of Science and Technology, VietNam
E-mail: bhthong@ute.udn.vn

Abstract. The cushions assistance to the pressure ulcers (PUs) prevention has like objective to improve wheelchair users by decreasing the effects of pressure between the both surfaces. Innovating honeycomb cushion (HC) containing thermoplastic polyurethane (TPU) which has many advantages was studied using a biomechanical and digital model for simulation. We used the finite element method (FEM) for the cushion simulation and modelization. This method led to the structural decomposition of the cushion in the aim to evaluate his mechanical behavior. This comportment was simulated numerically by FEM on cell of the cushion, then on the whole cushion. The tests of compression carried out with the electromechanical press (INSTRON 33R4204 machine) on HC permitted to determine curves related to his mechanical behavior. The result helps the comparison with the numerical model developed in this study. This result permits to study the distribution of pressure, shearing, friction and the microclimate which are sources of PUs through HC. The idea following this work is to optimize the HC in the aim to present a new product satisfying the needs for wheelchair users. The design of HC for prevention assistance of the scabs in only one layer is possible and present many advantages: optimization of the manufacturing process, improved ventilation, facilitated cleaning and decreasing cost.

1. Introduction

The pressure ulcers (PUs) is a localized injury to the skin and/or underlying tissue usually located over a bony prominence. In general, this lesion results from a pressure due to a compression effect and/or shear. Many contributing or confounding factors are also associated with pressure ulcers. It has been estimated that 90% of pressure ulcers occur at the contact between the buttock-tissue and the seat cushion. For people who use a wheelchair, the pressure ulcer often occurs in regions of the ischial, sacrum and the greater trochanter of the femur [1]. There are many ways to prevent pressure ulcer for improving seated comfort and reducing pressure ulcers for wheelchair users. The average cost of treatment was estimated in the United Kingdom (UK) according to the stages of disease severity. The ranges are from 1,200 € for a simple redness to 12,500 € for fourth stage that the depth of the wound is large and frequent infectious complications [2]. The pressure leads to pressure ulcers resulting in tissue hypoxia, which is defined as deprived of adequate oxygen supply in subcutaneous tissues. The reduction of blood supply and oxygen is mainly due to excessive pressure that would of blood vessels and leads to ischemia [3]. It is well-known that the movements appear mainly in the sagittal plane compared to the frontal plane [4]; a position change is required in spinal cord injury patients to relieve discomfort in subcutaneous tissues.

On the viewpoint of clinical medicine, the pressure ulcer mainly depends on several intrinsic and extrinsic factors. The intrinsic factors are attributed, among others, malnutrition, aging, obesity, incontinence, dry skin, muscles atrophy, poor posture and immobility. The extrinsic factors themselves are rather due to pressure, shear, friction and microclimate of the skin [1, 5, 6]. These extrinsic factors are the most important factor [7] and return to the patient's comfort and the idea of a welfare related to a particular seat position [8]. The comfort is a personal and subjective notion. It is comfortable for one, will not be necessarily for another. For example, people with smaller muscles, less resistant to deformation of tissue and it is easily, therefore, to be discomfort and tissue damage [9]. The weight is obviously considered and the Body Mass Index (BMI) of a person is calculated are greater than the increase of the pressure on the cushion [10-11]. The pressure distribution on cushions reduces the pressure ulcers. These studies were performed on paraplegic subjects, neurological, elderly and disabled, to observe the influence of the medium pressure. The high pressure, or the lower contact surface, while paraplegics have high-pressure peaks more than valid [12]. Furthermore, other studies to evaluate the relationship between the perception of seating comfort on a cushion and the interface pressure, the temperature and humidity have been conducted on wheelchair users. The results showed that the comfort was not necessarily related to the interface pressures, weak or important.
In addition, the discomfort was not necessarily related to the highest temperatures [7, 13]. It should be noted that users seem to privilege with a medium-firm level of cushion properties. The time factor strongly influences the feeling of comfort and can leads to fatigue. One static position for many hours, compress the tissues of the buttocks area and will be perceived as very unpleasant [9]. There are several interacting factors influence to the comfort [9]: (i) poor distribution of stresses in the soft tissue, (ii) the accumulation of moisture, (iii) the accumulation or heat loss, (iv) less stability. The aim of this study is to investigate the mechanical behavior of seat in a wheelchair, especially on a Stimulite cushion thermoplastic polyurethane honeycomb (TPU). The research presents many advantages compared to other previous studies.

2. Cushion Modeling

2.1. Spatial Geometry and CAD model

The objective is to model and numerically simulate the mechanical behavior of a Stimulite honeycomb cushion composed of thousands of flexible cells (Figure 1) and these numerical results from simulation are compared with those of experimental tests.

a) Flat Stimulite Cushion Classic

b) Structure of a honeycomb layer

Figure 1. Stimulite honeycomb cushion.

The cushion used in this study is made from an elastomers material TPU (Thermoplastic Polyurethane), non-toxic and safe for the people health and the environment, this material has good thermo-mechanical behavior, high compressive resistant, and puncture, tear and abrasion [14-15]. The flexible cellular TPU matrix is composed of preformed films in front of and heat-sealed in back side (Figure 2a). The alveolar films contain holes with diameter of 1.3 mm. Finally, the obtained matrix has two thicknesses, 0.2 mm and 0.4 mm on the non-heat sealed plans and the assembly planes, respectively (Figure 2b). The distance between two parallel faces is in the order of 7 mm.

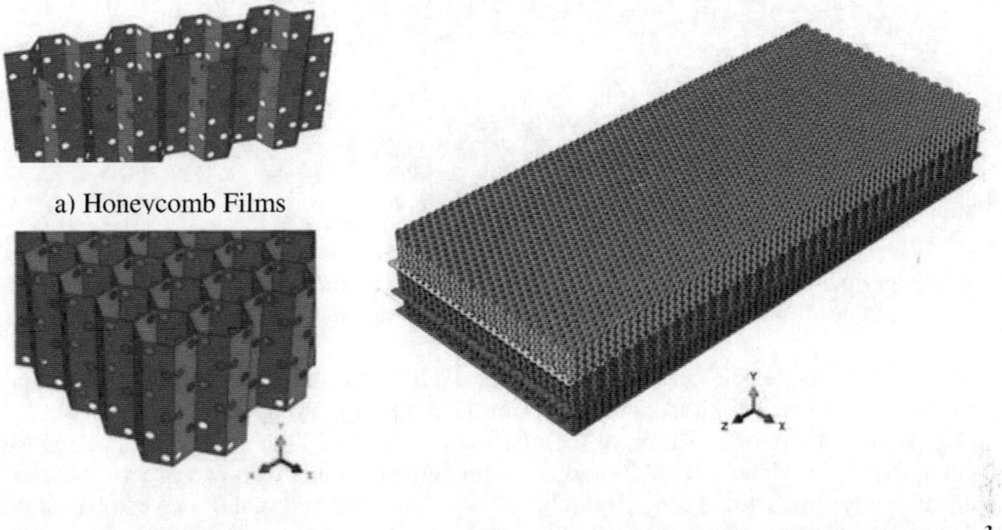

a) Honeycomb Films

b) Thermo-welded Honeycomb

c) Half-Honeycomb Cushion (**460x210x72 mm**^{3})

Figure 2. Thermal welding of the honeycomb cushion.

The CAD geometry of the half-honeycomb cushion was constructed by assembling 3 layers independent planes separated by SOLIDWORKS® software (Figure 2.c). The finite element modeling is created by commercial ABAQUS® software to investigate the effects of related pressure that appears at the contact interface buttocks/cushion. The honeycomb cushion, which was produced by Supracor® Company, made of thermoplastic polyurethane elastomer (TPU) with rate-independent hysteresis and viscoelastic behaviors. For isotropic hyperelastic materials (incompressibility material), the hyperelastic behaviors of the Mooney-Rivlin model are described by the strain energy potential function Eq. (1), W in the following form:

$$W = C_{10}(\bar{I}_1\text{-}3) + C_{01}(\bar{I}_2\text{-}3) + \frac{2}{D_1}(J-1)^2 \qquad (1)$$

Where $\bar{I}_1$, $\bar{I}_2$ are the first and the second invariant of the tensor, respectively, J is the Jacobine deformation. In Eq. (2) C_{10}, C_{01} and D_1 are the material-dependent parameters. $D_1 = \frac{1-2\nu}{C_{10}+C_{01}}$, where ν is the Poisson's ratio of the material. The material parameters of the honeycomb cushion (TPU) were obtained from tensile and compressive tests that were conducted on the INSTRON 33R4204 machine. The viscoelastic properties of the honeycomb cushion (TPU) are determined by plotting the normalized stress with initial stress σ_0 against the time. The viscoelastic model has been obtained from relaxation tests by fitting a three-term Prony series (Eq. 2) to the relaxation data as follows:

$$E(t) = 6.5(1 + 0.023e^{-t/12} + 0.08e^{-t/98} + 0.034e^{-t/1600}) \qquad (2)$$

The viscoelastic material parameters of the honeycomb cushion TPU were defined such that: N = 3; g_1 = 0.02 and τ_1 = 12 s; g_2 = 0.074 and τ_1 = 98 s, and g_3 = 0.031 and τ_3 = 1600 s.

2.2. Experimental Set-up

We performed an experimental of compression with a honeycomb layer (8 celles) (Figure 3.a) to validate the comportment of the TPU material on the machine compression ISTRON 33R4204 (Figure 3.b). The experimental compression tests were carried out with a speed of 0.1 mm/s and a maximum displacement imposed of 3 mm.

IOP Conf. Series: Materials Science and Engineering **540** (2019) 012015 doi:10.1088/1757-899X/540/1/012015

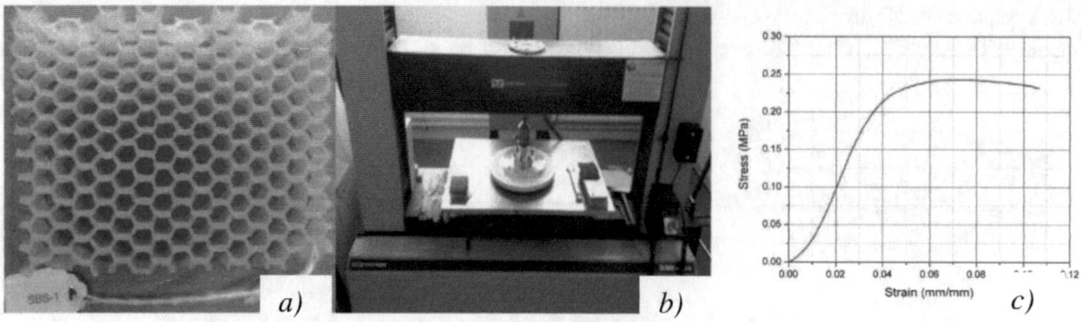

Figure 3. Experimental test of a honeycomb layer on the machine compression ISTRON 33R4204 and non-linear stress–strain curve of honeycomb cushion material (TPU).

Figure 3.c represents the tensile curve obtained from the tests for the TPU material. From this figure, some material parameters are extracted and used for the model according to Kanyanta et al., [16]. The hyperelastic parameters Mooney-Rivlin of the honeycomb cushion TPU as flow: C_{10} = 1.24 Mpa; C_{01} = 0.01 Mpa; D_1 = 2 (Mpa^{-1}). In this study, all numerical simulations were performed with the parameters of the hyper-elastic Money-Rivlin behavior of the TPU material (honeycomb cushion).
An experimental device was designed and fabricated in order to compress a TPU honeycomb cushion with a human buttocks prototype created from the 3D model meets the requirements of the ISO 16840-2 standard [17] (Figure 4.a). This standard is set to the physical and mechanical characteristics of the seat cushion in order to measure the displacement that generated by a vertical load applied on the planar surface of the honeycomb cushion, at a distance of 127 mm from the rear edge of the buttocks (Figure 4.a). The compression tests were performed on the "Stimulite Cushion Classic", with an electromechanical pressure Instron 33R4204 (50 KN load cell) (Figure 4.b). A displacement is imposed on the cushion so that the maximum achieved load of 830 N and maintained through the use of a PID (Proportional Integral Derivative) controller (Figure 4.b).

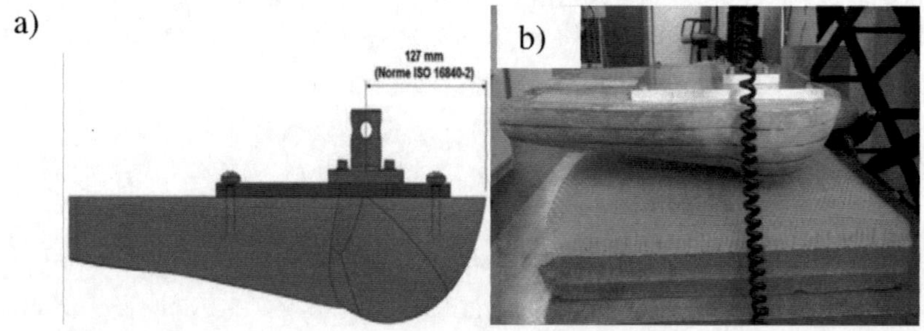

Figure 4. Design and implementation of prototype experimental model rigid buttocks, temporal compression tests on Stimulite honeycomb cushion.

The compression tests with electromechanical pressure the Stimulite cushion were identified temporal force curve related to the mechanical behaviors of the cushion. To achieve the maximum load of 830 N, a displacement of 41.14 mm was applied on the buttock at the load set point to 127 mm.

2.3. Numerical simulation of contact problems

To compare experimental and numerical results and validate the comportment of the honeycomb cushiopn (TPU material). We proposed numerical model, a honeycomb layer of cushion model, with about 8 celles on each side and a height of h = 20 mm (Figure 5.a), the explicit dynamic approach (DE) (ABAQUS® software) was used to simulate our problematic.

IOP Conf. Series: Materials Science and Engineering **540** (2019) 012015 doi:10.1088/1757-899X/540/1/012015

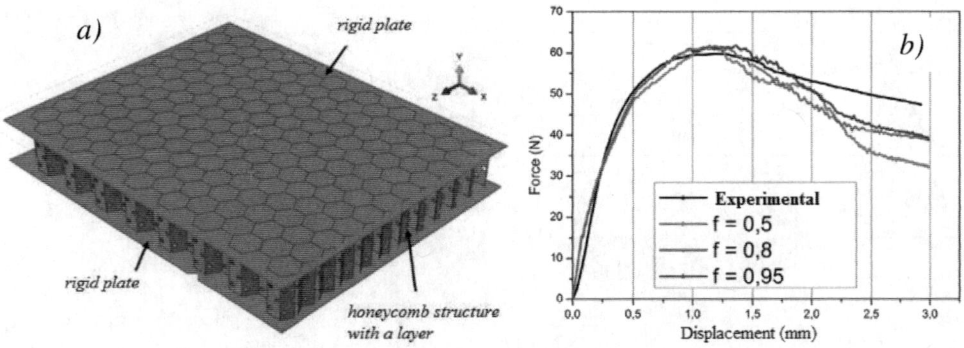

Figure 5. The geometry of a honeycomb layer and comparaison numerical simulation – experimental non-linear force – displacement curve of a honeycomb layer material (TPU).

Taking into account the interaction between the honeycomb layer with the two rigid plates was complicated. Indeed, we use to simulate the contact between the rigid plate and the honeycomb layer contact of a "surface-to-surface" type it possible to have numerical results much closer to the experimental curve (Figure 5.b) as well as a coherent deformation of the cells. The calculation time was 30 minutes with PC 16 processors of 32 Gb RAM. This approach allows the deformable elements in contact with the two rigid plates to under go large deformations without "crossing" the rigid elements. Then we wanted to study the influence of the coulomb friction coefficient between these plates and the cells. Comparing the numerical curves obtained for three different coefficients of friction (f = 0.5, 0.8 and 0.95) shows that the higher the coefficient of friction, the closer the numerical and experimental curves are (Figure 5.b), the contact so is almost sticky.

3. Numerical simulation applied to the honeycomb cushion

To reduce the calculation time and due to symmetries existing in the structure, only a half of rigid-buttock/honeycomb cushion model needs to be meshed and calculated. In this configuration, the model contains about that create a large mesh with a huge number of elements (Table 1).

Table 1. Characteristics of the mesh half-buttock/honeycomb cushion model.

Mesh data			
Type of elements	Type of elements	Type of elements	Type of elements
Honeycomb cushion	S3R	8876760	4788759
Rigid half-buttock	R3D3	42532	21268
Rigid plate	R3D3	80	55
	Total	**8919372**	**4810082**

The nodes of the mesh (Figure 6.b) located at the bottom surface (plane Y = 0) are encased (translations and zero rotations). To apply in the symmetry of the model, the nodes located in the symmetry plane (plane X = 0) are blocked in the X-direction translation and Y and Z-directions rotation. The honeycomb cushion used in this numerical simulation was only a half honeycomb cushion of 72 mm height. In addition, a half rigid-buttock, supposed in dimensionally stable, is requested with the application of a reduced dead weight of 41.5 kg applied in the vertical direction (Y).

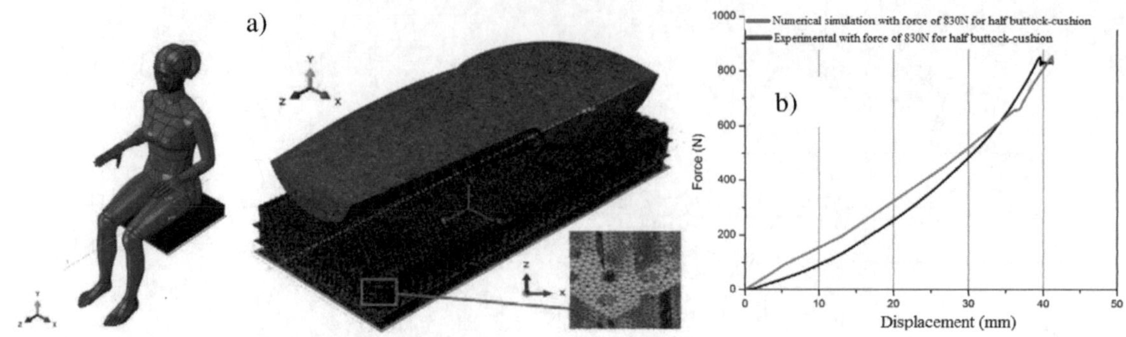

Figure 6. a) The geometry human body/cushion and mesh of a half-buttock/cushion model; b) The force - displacement curve obtained from the experimental tests and numerical simulation for the honeycomnb cushion.

In this configuration the calculation lasts approximately 48 hours 10 minutes with a PC 16 processors of 32 Gb of RAM. The vertical deformation (Y axis) of the cushion is shown in Figure 7.a. One can look at the maximum value of displacement on the honeycomb cushion in the vertical axis (Uy) of 41,33mm. The force – displacement curve compared with the experimental result above (compression tests on the traction machine (Figure 6.b)).

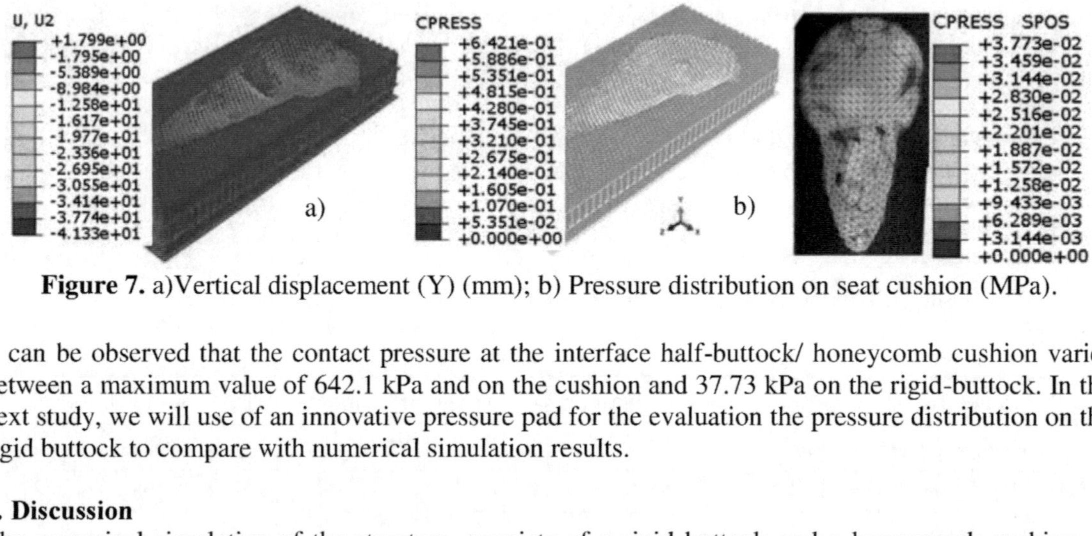

Figure 7. a)Vertical displacement (Y) (mm); b) Pressure distribution on seat cushion (MPa).

It can be observed that the contact pressure at the interface half-buttock/ honeycomb cushion varies between a maximum value of 642.1 kPa and on the cushion and 37.73 kPa on the rigid-buttock. In the next study, we will use of an innovative pressure pad for the evaluation the pressure distribution on the rigid buttock to compare with numerical simulation results.

4. Discussion

The numerical simulation of the structure, consists of a rigid buttock and a honeycomb cushion, is complicated and long computationally-intensive because it includes thousands of perforated cells that require very fine mesh. The structural decomposition of Stimulite cushion honeycomb allowed initially for focusing on the model with the isolated case of a cell to better understand their behaviors and to extend with many cases of alveolar parallelepiped volumes with approximately thousands of alveoli. The inspection of internal mechanical phenomena, such as stress and strain, revealed that it was possible to vary the operating parameters like the wall section, horizontal sheets addition, the position and diameter of the holes to make it less rigid the seat structure. The comfort depends upon the rigidity of the material, which ensures the strength and stability of the cushion. Hence, the finite element simulations can contribute to optimize the seating comfort by acting on this rigidity.

The proposed model in this paper is to model a cellular layer and it must be extended to multiple layers with different rigidity and mechanical properties depend on the type of TPU. The completed model will then study to obtain the impact of several geometrical parameters (diameter of the holes, sections, dimensions,...) on the mechanical behavior of the TPU cushion. Moreover, the optimal

IOP Conf. Series: Materials Science and Engineering **540** (2019) 012015 doi:10.1088/1757-899X/540/1/012015

solution will provide a cushion to the homogenized properties thereby decreasing the calculation time and will reduce manufacturing costs: optimized production, optimal ventilation, easy cleaning, cost …
The study of microclimate intrinsic factor (thermo-mechanical behavior) by numerical simulation is also possible. Thermodynamics is a study to be taken into account in the development of this project over the perforated foam composition of the cushion. It has to be simulated, and checked, after seating, if the interface temperature is similar to the experimental results. Thermodynamics must also take advantage of the cushion. In addition, the model should be adapted in order to simulate the air flow through the layers of the cushion, whether the breakdown of Stimulite cushion is sufficient to avoid maceration effects. Most car seats are made of polyurethane foam. Industrial automotive, aerospace, furniture should be focused on these cellular TPU solutions that seem most appropriate to the fluidic effects.

5. Conclusion

The majority of studies deal with mainly pressure distribution, but study of the mechanical behavior of cushions is considered to prevent bedsores. It is possible to stand out from this research that primarily focused on the pressure. Other phenomena are considered. The thermo-fluid-mechanical coupling at the numerical simulation, which should innovate by using a reverse lookup. The idea may be, then, performed this test with a model diaper non-rigid biomechanics using a law close to human reality behavior. This first study, the mechanical behavior of honeycomb cushion helps to highlight the importance of using an effective method to improve the design of cushions to serve as an aid in preventing pressure ulcers: it is the numerical simulation using the finite element method.

6. References

[1] Ewers B J 2004 *CA: Exp onent Failure Analysis Associates, Menlo Park.* Investigation of supracor stimulite wheelchair cushions.

[2] Bennett G 2004 The cost of pressure ulcers in the UK *Age and Ageing.* **33(3)** 230-35

[3] Le site de référence de l'escarre. *Tout ce qui traite de l'escarre,* http://www.escarre.fr/

[4] Reenalda J, Van Geffen P, Nederhand M, Jannink M, IJzerman M, and Rietman H 2009 Analysis of healthy sitting behavior: Interface pressure distribution and subcutaneous tissue oxygenation. *J. Rehabil. Res. Dev.* **46(5)** 577-86

[5] Barbenel J C 1991 Pressure management *Prosthetics and Orthotics International.* **15** 225 231.

[6] Bedard-Phane E 1999 Design de Coussins de Fauteuil Roulant à L'Aide d'un Modèle Biomécanique. *Mémoire maitriseès sciences appliquées, Ecole polytechnique de Montréal.*

[7] Dussault F P 2004 Les Aides Techniques à la Posture. (ATP Québec)

[8] Staarink H A M 1995. Sitting posture, Comfort and Pressure: assessing the quality of wheelchair cushions. *Delft University Press*

[9] Ferguson-Pell M 1990 Seat cushion selection *Journal of Rehabilitation Research & Development.* **27(2)** 49-73

[10] Shechtman O, 2001. Comparing wheelchair cushions for effectiveness of pressure relief: A pilot study *Occupational Therapy Journal of Research.* **21(1)** 1-14.

[11] Portnoy S, Vuillerme N, Payan Y, Gefen A 2011. Clinically oriented real-time monitoring of the individual's risk for deep tissue injury *Medical & Biological Engineering & Computing.* **49(4)** 473-83.

[12] Linder-Ganz E, Yarnitzky G, Yizhar Z, Siev-Ner I, Gefen A 2009 Real-Time Finite Element Monitoring of Sub-Dermal Tissue Stresses in Individuals with Spinal Cord Injury: Toward Prevention of Pressure Ulcers *Annals of Biomedical Engineering.* **37(2)** 387-400.

[13] Stockton L and Rithalia S 2009 Pressure-reducing cushions: Perceptions of comfort from the wheelchair users' perspective using interface pressure, temperature and humidity measurements *Journal of Tissue Viability.* **18(2)** 28-35.

[14] Biron M 2000 Elastomères Thermoplastiques TPE (AM3400) *Techniques de l'ingénieur.*

[15] Collette D 2000 Feuilles et films de polyuréthane thermoplastique (TPU) *Techniques de l'ingénieur.*

[16] Kanyanta V and Ivankovic A 2010 Mechanical characterisation of polyurethane elastomer for biomedical applications, *J. Mech. Behav. Biomed. Mater.* **3(1)** 51-62
[17] International standard ISO 16840-2 ISO, 2007. Wheelchair seating – Part 2: Determination of physical and mechanical characteristics of devices intended to manage tissue integrity – Seat cushions, available at: https://infostore.saiglobal.com/store/PreviewDoc.aspx?saleItemID=289364.

Acknowledgments
This research is funded by Funds for Science and Technology Development of the University of Danang under project number 46/HĐ-KHCN-2018 (B2018-ĐN06-12).

Institute of Physics
Dirac House, Temple Back
Bristol BS1 6BE UK

ISSN: 1757-8981
ISBN 978-1-5108-8920-0